FREE Test Taking Tips DVD Offer

To help us better serve you, we have developed a Test Taking Tips DVD that we would like to give you for FREE. **This DVD covers world-class test taking tips that you can use to be even more successful when you are taking your test.**

All that we ask is that you email us your feedback about your study guide. Please let us know what you thought about it – whether that is good, bad or indifferent.

To get your **FREE Test Taking Tips DVD**, email freedvd@studyguideteam.com with "FREE DVD" in the subject line and the following information in the body of the email:

a. The title of your study guide.

b. Your product rating on a scale of 1-5, with 5 being the highest rating.

c. Your feedback about the study guide. What did you think of it?

d. Your full name and shipping address to send your free DVD.

If you have any questions or concerns, please don't hesitate to contact us at freedvd@studyguideteam.com.

Thanks again!

GRE Math

WORKBOOK

GRE Math Prep 2018 & 2019 and Practice Tests for the Quantitative Reasoning Section of the GRE Exam

- Comprehensive Reviews
- Proven Test Strategies
- Practice Test Questions

Test Prep Books

Two Complete Practice Tests

GRE Math Workbook

GRE Quantitative Reasoning Practice Team

Table of Contents

Quick Overview

As you draw closer to taking your exam, effective preparation becomes more and more important. Thankfully, you have this study guide to help you get ready. Use this guide to help keep your studying on track and refer to it often.

This study guide contains several key sections that will help you be successful on your exam. The guide contains tips for what you should do the night before and the day of the test. Also included are test-taking tips. Knowing the right information is not always enough. Many well-prepared test takers struggle with exams. These tips will help equip you to accurately read, assess, and answer test questions.

A large part of the guide is devoted to showing you what content to expect on the exam and to helping you better understand that content. Near the end of this guide is a practice test so that you can see how well you have grasped the content. Then, answer explanations are provided so that you can understand why you missed certain questions.

Don't try to cram the night before you take your exam. This is not a wise strategy for a few reasons. First, your retention of the information will be low. Your time would be better used by reviewing information you already know rather than trying to learn a lot of new information. Second, you will likely become stressed as you try to gain a large amount of knowledge in a short amount of time. Third, you will be depriving yourself of sleep. So be sure to go to bed at a reasonable time the night before. Being well-rested helps you focus and remain calm.

Be sure to eat a substantial breakfast the morning of the exam. If you are taking the exam in the afternoon, be sure to have a good lunch as well. Being hungry is distracting and can make it difficult to focus. You have hopefully spent lots of time preparing for the exam. Don't let an empty stomach get in the way of success!

When travelling to the testing center, leave earlier than needed. That way, you have a buffer in case you experience any delays. This will help you remain calm and will keep you from missing your appointment time at the testing center.

Be sure to pace yourself during the exam. Don't try to rush through the exam. There is no need to risk performing poorly on the exam just so you can leave the testing center early. Allow yourself to use all of the allotted time if needed.

Remain positive while taking the exam even if you feel like you are performing poorly. Thinking about the content you should have mastered will not help you perform better on the exam.

Once the exam is complete, take some time to relax. Even if you feel that you need to take the exam again, you will be well served by some down time before you begin studying again. It's often easier to convince yourself to study if you know that it will come with a reward!

Test-Taking Strategies

1. Predicting the Answer

When you feel confident in your preparation for a multiple-choice test, try predicting the answer before reading the answer choices. This is especially useful on questions that test objective factual knowledge or that ask you to fill in a blank. By predicting the answer before reading the available choices, you eliminate the possibility that you will be distracted or led astray by an incorrect answer choice. You will feel more confident in your selection if you read the question, predict the answer, and then find your prediction among the answer choices. After using this strategy, be sure to still read all of the answer choices carefully and completely. If you feel unprepared, you should not attempt to predict the answers. This would be a waste of time and an opportunity for your mind to wander in the wrong direction.

2. Reading the Whole Question

Too often, test takers scan a multiple-choice question, recognize a few familiar words, and immediately jump to the answer choices. Test authors are aware of this common impatience, and they will sometimes prey upon it. For instance, a test author might subtly turn the question into a negative, or he or she might redirect the focus of the question right at the end. The only way to avoid falling into these traps is to read the entirety of the question carefully before reading the answer choices.

3. Looking for Wrong Answers

Long and complicated multiple-choice questions can be intimidating. One way to simplify a difficult multiple-choice question is to eliminate all of the answer choices that are clearly wrong. In most sets of answers, there will be at least one selection that can be dismissed right away. If the test is administered on paper, the test taker could draw a line through it to indicate that it may be ignored; otherwise, the test taker will have to perform this operation mentally or on scratch paper. In either case, once the obviously incorrect answers have been eliminated, the remaining choices may be considered. Sometimes identifying the clearly wrong answers will give the test taker some information about the correct answer. For instance, if one of the remaining answer choices is a direct opposite of one of the eliminated answer choices, it may well be the correct answer. The opposite of obviously wrong is obviously right! Of course, this is not always the case. Some answers are obviously incorrect simply because they are irrelevant to the question being asked. Still, identifying and eliminating some incorrect answer choices is a good way to simplify a multiple-choice question.

4. Don't Overanalyze

Anxious test takers often overanalyze questions. When you are nervous, your brain will often run wild, causing you to make associations and discover clues that don't actually exist. If you feel that this may be a problem for you, do whatever you can to slow down during the test. Try taking a deep breath or counting to ten. As you read and consider the question, restrict yourself to the particular words used by the author. Avoid thought tangents about what the author *really* meant, or what he or she was *trying* to say. The only things that matter on a multiple-choice test are the words that are actually in the question. You must avoid reading too much into a multiple-choice question, or supposing that the writer meant something other than what he or she wrote.

5. No Need for Panic

It is wise to learn as many strategies as possible before taking a multiple-choice test, but it is likely that you will come across a few questions for which you simply don't know the answer. In this situation, avoid panicking. Because most multiple-choice tests include dozens of questions, the relative value of a single wrong answer is small. Moreover, your failure on one question has no effect on your success elsewhere on the test. As much as possible, you should compartmentalize each question on a multiple-choice test. In other words, you should not allow your feelings about one question to affect your success on the others. When you find a question that you either don't understand or don't know how to answer, just take a deep breath and do your best. Read the entire question slowly and carefully. Try rephrasing the question a couple of different ways. Then, read all of the answer choices carefully. After eliminating obviously wrong answers, make a selection and move on to the next question.

6. Confusing Answer Choices

When working on a difficult multiple-choice question, there may be a tendency to focus on the answer choices that are the easiest to understand. Many people, whether consciously or not, gravitate to the answer choices that require the least concentration, knowledge, and memory. This is a mistake. When you come across an answer choice that is confusing, you should give it extra attention. A question might be confusing because you do not know the subject matter to which it refers. If this is the case, don't eliminate the answer before you have affirmatively settled on another. When you come across an answer choice of this type, set it aside as you look at the remaining choices. If you can confidently assert that one of the other choices is correct, you can leave the confusing answer aside. Otherwise, you will need to take a moment to try to better understand the confusing answer choice. Rephrasing is one way to tease out the sense of a confusing answer choice.

7. Your First Instinct

Many people struggle with multiple-choice tests because they overthink the questions. If you have studied sufficiently for the test, you should be prepared to trust your first instinct once you have carefully and completely read the question and all of the answer choices. There is a great deal of research suggesting that the mind can come to the correct conclusion very quickly once it has obtained all of the relevant information. At times, it may seem to you as if your intuition is working faster even than your reasoning mind. This may in fact be true. The knowledge you obtain while studying may be retrieved from your subconscious before you have a chance to work out the associations that support it. Verify your instinct by working out the reasons that it should be trusted.

8. Key Words

Many test takers struggle with multiple-choice questions because they have poor reading comprehension skills. Quickly reading and understanding a multiple-choice question requires a mixture of skill and experience. To help with this, try jotting down a few key words and phrases on a piece of scrap paper. Doing this concentrates the process of reading and forces the mind to weigh the relative importance of the question's parts. In selecting words and phrases to write down, the test taker thinks about the question more deeply and carefully. This is especially true for multiple-choice questions that are preceded by a long prompt.

9. Subtle Negatives

One of the oldest tricks in the multiple-choice test writer's book is to subtly reverse the meaning of a question with a word like *not* or *except*. If you are not paying attention to each word in the question, you can easily be led astray by this trick. For instance, a common question format is, "Which of the following is...?" Obviously, if the question instead is, "Which of the following is not...?," then the answer will be quite different. Even worse, the test makers are aware of the potential for this mistake and will include one answer choice that would be correct if the question were not negated or reversed. A test taker who misses the reversal will find what he or she believes to be a correct answer and will be so confident that he or she will fail to reread the question and discover the original error. The only way to avoid this is to practice a wide variety of multiple-choice questions and to pay close attention to each and every word.

10. Reading Every Answer Choice

It may seem obvious, but you should always read every one of the answer choices! Too many test takers fall into the habit of scanning the question and assuming that they understand the question because they recognize a few key words. From there, they pick the first answer choice that answers the question they believe they have read. Test takers who read all of the answer choices might discover that one of the latter answer choices is actually *more* correct. Moreover, reading all of the answer choices can remind you of facts related to the question that can help you arrive at the correct answer. Sometimes, a misstatement or incorrect detail in one of the latter answer choices will trigger your memory of the subject and will enable you to find the right answer. Failing to read all of the answer choices is like not reading all of the items on a restaurant menu: you might miss out on the perfect choice.

11. Spot the Hedges

One of the keys to success on multiple-choice tests is paying close attention to every word. This is never more true than with words like *almost*, *most*, *some*, and *sometimes*. These words are called "hedges" because they indicate that a statement is not totally true or not true in every place and time. An absolute statement will contain no hedges, but in many subjects, like literature and history, the answers are not always straightforward or absolute. There are always exceptions to the rules in these subjects. For this reason, you should favor those multiple-choice questions that contain hedging language. The presence of qualifying words indicates that the author is taking special care with his or her words, which is certainly important when composing the right answer. After all, there are many ways to be wrong, but there is only one way to be right! For this reason, it is wise to avoid answers that are absolute when taking a multiple-choice test. An absolute answer is one that says things are either all one way or all another. They often include words like *every*, *always*, *best*, and *never*. If you are taking a multiple-choice test in a subject that doesn't lend itself to absolute answers, be on your guard if you see any of these words.

12. Long Answers

In many subject areas, the answers are not simple. As already mentioned, the right answer often requires hedges. Another common feature of the answers to a complex or subjective question are qualifying clauses, which are groups of words that subtly modify the meaning of the sentence. If the question or answer choice describes a rule to which there are exceptions or the subject matter is complicated, ambiguous, or confusing, the correct answer will require many words in order to be expressed clearly and accurately. In essence, you should not be deterred by answer choices that seem excessively long. Oftentimes, the author of the text will not be able to write the correct answer without offering some qualifications and modifications. Your job is to read the answer choices thoroughly and completely and to select the one that most accurately and precisely answers the question.

13. Restating to Understand

Sometimes, a question on a multiple-choice test is difficult not because of what it asks but because of how it is written. If this is the case, restate the question or answer choice in different words. This process serves a couple of important purposes. First, it forces you to concentrate on the core of the question. In order to rephrase the question accurately, you have to understand it well. Rephrasing the question will concentrate your mind on the key words and ideas. Second, it will present the information to your mind in a fresh way. This process may trigger your memory and render some useful scrap of information picked up while studying.

14. True Statements

Sometimes an answer choice will be true in itself, but it does not answer the question. This is one of the main reasons why it is essential to read the question carefully and completely before proceeding to the answer choices. Too often, test takers skip ahead to the answer choices and look for true statements. Having found one of these, they are content to select it without reference to the question above. Obviously, this provides an easy way for test makers to play tricks. The savvy test taker will always read the entire question before turning to the answer choices. Then, having settled on a correct answer choice, he or she will refer to the original question and ensure that the selected answer is relevant. The mistake of choosing a correct-but-irrelevant answer choice is especially common on questions related to specific pieces of objective knowledge, like historical or scientific facts. A prepared test taker will have a wealth of factual knowledge at his or her disposal, and should not be careless in its application.

15. No Patterns

One of the more dangerous ideas that circulates about multiple-choice tests is that the correct answers tend to fall into patterns. These erroneous ideas range from a belief that B and C are the most common right answers, to the idea that an unprepared test-taker should answer "A-B-A-C-A-D-A-B-A." It cannot be emphasized enough that pattern-seeking of this type is exactly the WRONG way to approach a multiple-choice test. To begin with, it is highly unlikely that the test maker will plot the correct answers according to some predetermined pattern. The questions are scrambled and delivered in a random order. Furthermore, even if the test maker was following a pattern in the assignation of correct answers, there is no reason why the test taker would know which pattern he or she was using. Any attempt to discern a pattern in the answer choices is a waste of time and a distraction from the real work of taking the test. A test taker would be much better served by extra preparation before the test than by reliance on a pattern in the answers.

FREE DVD OFFER

Don't forget that doing well on your exam includes both understanding the test content and understanding how to use what you know to do well on the test. We offer a completely FREE Test Taking Tips DVD that covers world class test taking tips that you can use to be even more successful when you are taking your test.

All that we ask is that you email us your feedback about your study guide. To get your **FREE Test Taking Tips DVD**, email freedvd@studyguideteam.com with "FREE DVD" in the subject line and the following information in the body of the email:

- The title of your study guide.
- Your product rating on a scale of 1-5, with 5 being the highest rating.
- Your feedback about the study guide. What did you think of it?
- Your full name and shipping address to send your free DVD.

Introduction to the GRE

Function of the Test

The Graduate Record Examination (GRE) General Test is a standardized test administered by the Educational Testing Service (ETS) and used as part of the admissions process by masters, doctoral, and business programs at various universities. Specifically, the test is accepted or required by virtually every graduate and business program in the United States, as well as many schools around the world. There are also seven GRE tests in specific subject areas, but "GRE" in common usage refers only to the General Test.

In recent years, around 500,000 people have taken the GRE annually. Because the GRE is exclusively used as an admissions exam, most test takers are seniors in undergraduate programs who are planning to attend graduate school or college graduates who are seeking a graduate degree.

Test Administration

In the United States, the GRE is administered by computer, year-round at Prometric testing centers and, from time to time, on specific dates at other testing centers. In the typical middle-sized city, a location to take the test will be available somewhere in town on most days of any given month. Outside the U.S., the GRE is administered by computer or, where computer-testing sites are not available, by paper.

The fee for taking the GRE is the same worldwide, with the exception of China, where fees are slightly higher. The computer-based version of the test can be taken up to five times within any rolling twelve-month window, although test takers must wait at least 21 days after an attempt to retake the exam. Note, however, that individual schools' rules about how they treat retest scores may vary. Reasonable accommodations are available for test takers with disabilities, provided requests are submitted to, and approved by, ETS before the test taker schedules a test date. Requests may be made through the test taker's electronic account with ETS.

Test Format

The computer-based GRE is "adaptive by section," meaning that the difficulty of the second verbal and second quantitative sections that the test taker receives will depend on the his or her performance on the first of such sections completed. Test takers will have access to an on-screen calculator and may not use one of their own. Some questions provide multiple-choice answers, while others require test takers to fill-in-the-blank.

The total test time is around 3 hours and 45 minutes, broken down as follows:

Section	Time	Description
Analytical Writing	60 minutes	Two 30-minute "issue" and "argument" writing tasks
Verbal Reasoning	2 30-minute sections	20 questions assessing reading comprehension, critical reasoning, and vocabulary usage
Quantitative Reasoning	2 35-minute sections	20 questions assessing quantitative comparisons, problem solving items, and data interpretation questions
Experimental/Research Section	1 30- or 35-minute unscored section	May be either Verbal Reasoning or Quantitative Reasoning

Scoring

On the Verbal and Quantitative Reasoning sections, students receive a raw score that is simply the total number of questions answered correctly. There is no penalty for guessing. The raw score is then scaled to a score that ranges from 130 to 170. This score is available upon completion of the test.

There is no set "passing" score on the GRE; rather, each school considers test takers' scores relative to the school's standards and to the scores of other applicants. The average score on the test is between 150 and 152, while average scores for applicants for elite programs might be around 160.

The writing sections are scored later, on a scale from 0 to 6 in 0.5 point increments. The average score is around a 3.5.

Recent and Future Developments

The GRE has undergone major revisions over the years, most recently with the introduction of the current "GRE revised" test in 2011. Prior to that revision, the test was adaptive from question to question (rather than from section to section) and was scored on a 1600-point scale. No substantial changes have been announced recently.

Quantitative Reasoning

Arithmetic

Integers

An integer is any number that has a nonzero decimal part. This includes all positive and negative whole numbers and zero. Fractions and decimals—which aren't whole numbers—aren't integers.

Prime Numbers

A *prime* number cannot be divided except by 1 and itself. A prime number has no other factors, which means that no other combination of whole numbers can be multiplied to reach that number. For example, the set of prime numbers between 1 and 27 is {2, 3, 5, 7, 11, 13, 17, 19, 23}.

The number 7 is a prime number because its only factors are 1 and 7. In contrast, 12 isn't a prime number, as it can be divided by other numbers like 2, 3, 4, and 6. Because of they are composed of multiple factors, numbers like 12 are called *composite* numbers. All numbers greater than 1 that aren't prime numbers are composite numbers.

Even and Odd Numbers

An integer is *even* if one of its factors is 2, while those integers without a factor of 2 are *odd*. No numbers except for integers can have either of these labels. For example, 2, 40, -16, and 108 are all even numbers, while -1, 13, 59, and 77 are all odd numbers since they are integers that cannot be divided by 2 without a remainder. Numbers like 0.4, $\frac{5}{9}$, π, and $\sqrt{7}$ are neither odd nor even because they are not integers.

Decimals

A decimal number is designated by a decimal point, which indicates that what follows the point is a value that is less than 1 and is added to the integer number preceding the decimal point. The digit immediately following the decimal point is in the tenths place, the digit following the tenths place is in the hundredths place, and so on.

For example, the decimal number 1.735 has a value greater than 1 but less than 2. The 7 represents seven-tenths of the unit 1 (0.7 or $\frac{7}{10}$); the 3 represents three-hundredths of 1 (0.03 or $\frac{3}{100}$); and the 5 represents five-thousandths of 1 (0.005 or $\frac{5}{1000}$).

Rational and Irrational Numbers

Rational numbers include all numbers that can be expressed as a fraction; in other words, rational numbers encompass all integers and all numbers with terminating or repeating decimals. That is, any rational number either will have a countable number of nonzero digits or will end with an ellipses or a bar (3.6666... or $3.\overline{6}$) to depict repeating decimal digits. Some examples of rational numbers include 12, -3.54, $110.\overline{256}$, $\frac{-35}{10}$, and $4.\overline{7}$.

Irrational numbers include all real numbers that aren't rational. Irrational numbers can be thought of as any number with endless non-repeating digits to the right of the decimal point. They can be expressed as an endless decimal but never as a fraction. The most common irrational number is π, which has an endless and non-repeating decimal, but there are other well-known irrational numbers like *e* and $\sqrt{2}$.

Real Numbers

Defined by Descartes in the seventeenth century, real numbers include all numbers found on an infinite number line. All irrational and rational numbers are real numbers. Non-terminating decimal numbers and π are also real numbers. As the range of real numbers extends to both negative and positive infinity, the set of real numbers is complete and uncountable. This set is known as the complete ordered field of numbers.

Addition

Addition is the combination of two numbers so their quantities are added together cumulatively. The sign for an addition operation is the + symbol. For example, 9 + 6 = 15. The 9 and 6 combine to achieve a cumulative value, called a *sum*.

Addition holds the *commutative property*, which means that numbers in an addition equation can be switched without altering the result. The formula for the commutative property is a + b = b + a. Let's look at a few examples to see how the commutative property works:

$$7 = 3 + 4 = 4 + 3 = 7$$

$$20 = 12 + 8 = 8 + 12 = 20$$

Addition also holds the *associative property*, which means that the grouping of numbers doesn't matter in an addition problem. In other words, the presence or absence of parentheses is irrelevant. The formula for the associative property is (a + b) + c = a + (b + c). Here are some examples of the associative property at work:

$$30 = (6 + 14) + 10 = 6 + (14 + 10) = 30$$

$$35 = 8 + (2 + 25) = (8 + 2) + 25 = 35$$

Subtraction

Subtraction is taking away one number from another, so their quantities are reduced. The sign designating a subtraction operation is the – symbol, and the result is called the *difference*. For example, 9 - 6 = 3. The number *6* detracts from the number *9* to reach the difference *3*.

Unlike addition, subtraction follows neither the commutative nor associative properties. The order and grouping in subtraction impact the result.

$$15 = 22 - 7 \neq 7 - 22 = -15$$

$$3 = (10 - 5) - 2 \neq 10 - (5 - 2) = 7$$

When working through subtraction problems involving larger numbers, it's necessary to regroup the numbers. Let's work through a practice problem using regrouping:

$$\begin{array}{r} 325 \\ -\ 77 \\ \hline \end{array}$$

Here, it is clear that the ones and tens columns for 77 are greater than the ones and tens columns for 325. To subtract this number, borrow from the tens and hundreds columns. When borrowing from a column, subtracting 1 from the lender column will add 10 to the borrower column:

$$\begin{array}{rrr} 3-1 & 10+2-1 & 10+5 \\ - & 7 & 7 \\ \hline \end{array} = \begin{array}{rrr} 2 & 11 & 15 \\ - & 7 & 7 \\ \hline 2 & 4 & 8 \end{array}$$

After ensuring that each digit in the top row is greater than the digit in the corresponding bottom row, subtraction can proceed as normal, and the answer is found to be 248.

Multiplication

Multiplication involves adding together multiple copies of a number. It is indicated by an $\times$ symbol or a number immediately outside of a parenthesis, e.g. 5(8-2). The two numbers being multiplied together are called *factors*, and their result is called a *product*. For example, $9 \times 6 = 54$. This can be shown alternatively by expansion of either the 9 or the 6:

$$9 \times 6 = 9 + 9 + 9 + 9 + 9 + 9 = 54$$

$$9 \times 6 = 6 + 6 + 6 + 6 + 6 + 6 + 6 + 6 + 6 = 54$$

Like addition, multiplication holds the commutative and associative properties:

$$115 = 23 \times 5 = 5 \times 23 = 115$$

$$84 = 3 \times (7 \times 4) = (3 \times 7) \times 4 = 84$$

Multiplication also follows the distributive property, which allows the multiplication to be distributed through parentheses. The formula for distribution is $a \times (b + c) = ab + ac$. This is clear after the examples:

$$45 = 5 \times 9 = 5(3 + 6) = (5 \times 3) + (5 \times 6) = 15 + 30 = 45$$

$$20 = 4 \times 5 = 4(10 - 5) = (4 \times 10) - (4 \times 5) = 40 - 20 = 20$$

Multiplication becomes slightly more complicated when multiplying numbers with decimals. The easiest way to answer these problems is to ignore the decimals and multiply as if they were whole numbers. After multiplying the factors, place a decimal in the product. The placement of the decimal is determined by taking the cumulative number of decimal places in the factors.

For example:

$$\begin{array}{r} 0.7 \\ \times 3 \\ \hline 2.1 \end{array} \qquad \begin{array}{r} 2.6 \\ \times\ 4.2 \\ \hline 10.92 \end{array} \qquad \begin{array}{r} 1.5 \\ \times 6.4 \\ \hline 9.60 \end{array}$$

Let's tackle the first example. First, ignore the decimal and multiply the numbers as though they were whole numbers to arrive at a product: 21. Second, count the number of digits that follow a decimal (one). Finally, move the decimal place that many positions to the left, as the factors have only one decimal place. The second example works the same way, except that there are two total decimal places in the factors, so the product's decimal is moved two places over. In the third example, the decimal should be moved over two digits, but the digit zero is no longer needed, so it is erased and the final answer is 9.6.

Division

Division and multiplication are inverses of each other in the same way that addition and subtraction are opposites. The signs designating a division operation are the ÷ and / symbols. In division, the second number divides into the first.

The number before the division sign is called the *dividend* or, if expressed as a fraction, the *numerator*. For example, in $a \div b$, a is the dividend, while in $\frac{a}{b}$, a is the numerator.

The number after the division sign is called the *divisor* or, if expressed as a fraction, the *denominator*. For example, in $a \div b$, b is the divisor, while in $\frac{a}{b}$, b is the denominator.

Like subtraction, division doesn't follow the commutative property, as it matters which number comes before the division sign, and division doesn't follow the associative or distributive properties for the same reason. For example:

$$\frac{3}{2} = 9 \div 6 \neq 6 \div 9 = \frac{2}{3}$$

$$2 = 10 \div 5 = (30 \div 3) \div 5 \neq 30 \div (3 \div 5) = 30 \div \frac{3}{5} = 50$$

$$25 = 20 + 5 = (40 \div 2) + (40 \div 8) \neq 40 \div (2 + 8) = 40 \div 10 = 4$$

If a divisor doesn't divide into a dividend an integer-number of times, whatever is left over is termed the remainder. The remainder can be further divided out into decimal form by using long division; however, this doesn't always give a quotient with a finite number of decimal places, so the remainder can also be expressed as a fraction over the original divisor.

Division with decimals is similar to multiplication with decimals in that when dividing a decimal by a whole number, the decimal is ignored and the number is divided as if it were a whole number. Upon finding the answer, or quotient, the decimal is placed at the decimal place equal to that in the dividend.

$$15.75 \div 3 = 5.25$$

When the divisor is a decimal number, one needs to multiply both the divisor and dividend by 10 and repeat this until the divisor is a whole number. Then the division operation is completed as described above.

$$17.5 \div 2.5 = 175 \div 25 = 7$$

Factorization

Factors are the numbers multiplied to achieve a product. Thus, every product in a multiplication equation has, at minimum, two factors. Of course, some products will have more than two factors. For the sake of most discussions, one can assume that factors are positive integers.

To find a number's factors, one should start with 1 and the number itself. The next step is to divide the number by 2, 3, 4, and so on, to see if any divisors can divide the number without a remainder. A list should be kept of those that do. This process can be stopped upon reaching either the number itself or another factor.

For example, to find the factors of 45, the first step is to start with 1 and 45. The next step is to try to divide 45 by 2, which fails. After this, 45 gets divided by 3. The answer is 15, so 3 and 15 are now factors. Dividing by 4 doesn't work, and dividing by 5 leaves 9. Lastly, dividing 45 by 6, 7, and 8 all don't work. The next integer to try is 9, but this is already known to be a factor, so the factorization is complete. The factors of 45 are 1, 3, 5, 9, 15 and 45.

Prime Factorization

Prime factorization involves an additional step after breaking a number down to its factors: breaking down the factors until they are all prime numbers. A prime number is any number that can only be divided by 1 and itself. The prime numbers between 1 and 20 are 2, 3, 5, 7, 11, 13, 17, and 19. As a simple test, numbers that are even or end in 5 are not prime.

When attempting to break 129 down into its prime factors, the factors are found first: 3 and 43. Both 3 and 43 are prime numbers, so that means the prime factorization is complete. If 43 was not a prime number, then it would also need to be factorized until all of the factors were expressed as prime numbers.

Common Factor

A common factor is a factor shared by two numbers. The following examples demonstrate how to find the common factors of 45 and 30:

- The factors of 45 are: 1, 3, 5, 9, 15, and 45.
- The factors of 30 are: 1, 2, 3, 5, 6, 10, 15, and 30.
- The common factors are 1, 3, 5, and 15.

Greatest Common Factor

The greatest common factor is the largest number among the shared, common factors. From the factors of 45 and 30, the common factors are 3, 5, and 15. Thus, 15 is the greatest common factor, as it's the largest number.

Least Common Multiple

The least common multiple is the smallest number that's a multiple of two numbers. For example, to find the least common multiple of 4 and 9, the multiples of 4 and 9 are found first. The multiples of 4 are 4, 8, 12, 16, 20, 24, 28, 32, 36, and so on. For 9, the multiples are 9, 18, 27, 36, 45, 54, etc. Thus, the least common multiple of 4 and 9 is 36, the lowest number where 4 and 9 share multiples.

If two numbers share no factors besides 1 in common, then their least common multiple will be simply their product. If two numbers have common factors, then their least common multiple will be their product divided by their greatest common factor. This can be visualized by the formula $LCM = \frac{x \times y}{GCF}$, where *x* and *y* are some integers and *LCM* and *GCF* are their least common multiple and greatest common factor, respectively.

Exponents

An *exponent* is an operation used as shorthand for a number multiplied or divided by itself for a defined number of times.

$$3^7 = 3 \times 3 \times 3 \times 3 \times 3 \times 3 \times 3$$

In this example, the 3 is called the *base* and the 7 is called the *exponent*. The exponent is typically expressed as a superscript number near the upper right side of the base, but can also be identified as the number following a caret symbol (^). This operation is verbally expressed as "3 to the 7th power" or "3 raised to the power of 7." Common exponents are 2 and 3. A base raised to the power of 2 is referred to as having been "squared," while a base raised to the power of 3 is referred to as having been "cubed."

Several special rules apply to exponents. First, the *Zero Power Rule* finds that any number raised to the zero power equals 1. For example, 100^0, 2^0, $(-3)^0$and 0^0 all equal 1 because the bases are raised to the zero power.

Second, exponents can be negative. With negative exponents, the equation is expressed as a fraction, as in the following example:

$$3^{-7} = \frac{1}{3^7} = \frac{1}{3 \text{ x } 3 \text{ x } 3 \text{ x } 3 \text{ x } 3 \text{ x } 3 \text{ x } 3}$$

Third, the *Power Rule* concerns exponents being raised by another exponent. When this occurs, the exponents are multiplied by each other:

$$(x^2)^3 = x^6 = (x^3)^2$$

Fourth, when multiplying two exponents with the same base, the *Product Rule* requires that the base remains the same and the exponents are added. For example, a^xx a^y = a^{x+y}. Since addition and multiplication are commutative, the two terms being multiplied can be in any order.

$$x^3x^5 = x^{3+5} = x^8 = x^{5+3} = x^5x^3$$

Fifth, when dividing two exponents with the same base, the *Quotient Rule* requires that the base remains the same, but the exponents are subtracted. So, $a^x \div a^y = a^{x-y}$. Since subtraction and division are not commutative, the two terms must remain in order.

$$x^5x^{-3} = x^{5-3} = x^2 = x^5 \div x^3 = \frac{x^5}{x^3}$$

Additionally, 1 raised to any power is still equal to 1, and any number raised to the power of 1 is equal to itself. In other words, $a^1 = a$ and $14^1 = 14$.

Exponents play an important role in scientific notation to present extremely large or small numbers as follows: $a \times 10^b$. To write the number in scientific notation, the decimal is moved until there is only one digit on the left side of the decimal point, indicating that the number *a* has a value between 1 and 10. The number of times the decimal moves indicates the exponent to which 10 is raised, here represented by b. If the decimal moves to the left, then *b* is positive, but if the decimal moves to the right, then *b* is negative. The following examples demonstrate these concepts:

$$3{,}050 = 3.05 \times 10^3$$

$$-777 = -7.77 \times 10^2$$

$$0.000123 = 1.23 \times 10^{-4}$$

$$-0.0525 = -5.25 \times 10^{-2}$$

Roots

The *square root symbol* is expressed as $\sqrt{\ }$ and is commonly known as the *radical*. Taking the root of a number is the inverse operation of multiplying that number by itself some number of times. For example, squaring the number 7 is equal to 7 × 7, or 49. Finding the square root is the opposite of finding an exponent, as the operation seeks a number that when multiplied by itself, equals the number in the square root symbol.

For example, $\sqrt{36}$ = 6 because 6 multiplied by 6 equals 36. Note, the square root of 36 is also -6 since -6 × -6 = 36. This can be indicated using a plus/minus symbol like this: ±6. However, square roots are often just expressed as a positive number for simplicity, with it being understood that the true value can be either positive or negative.

Perfect squares are numbers with whole number square roots. The list of perfect squares begins with 0, 1, 4, 9, 16, 25, 36, 49, 64, 81, and 100.

Determining the square root of imperfect squares requires a calculator to reach an exact figure. It's possible, however, to approximate the answer by finding the two perfect squares that the number fits between. For example, the square root of 40 is between 6 and 7 since the squares of those numbers are 36 and 49, respectively.

Square roots are the most common root operation. If the radical doesn't have a number to the upper left of the symbol $\sqrt{\ }$, then it's a square root. Sometimes a radical includes a number in the upper left, like $\sqrt[3]{27}$, as in the other common root type—the cube root. Complicated roots, like the cube root, often require a calculator.

Percentages

Percentages can be thought of as fractions with a denominator of 100. In fact, percentage means "per hundred." Problems often require converting numbers from percentages, fractions, and decimals. The following explains how to work through those conversions.

Converting Fractions to Percentages: The fraction is converted by using an equivalent fraction with a denominator of 100. For example, $\frac{3}{4} = \frac{3}{4} \times \frac{25}{25} = \frac{75}{100} = 75\%$.

Converting Percentages to Fractions: Percentages can be converted to fractions by turning the percentage into a fraction with a denominator of 100. Test takers should be wary of questions asking the converted fraction to be written in the simplest form. For example, $35\% = \frac{35}{100}$ which, although correctly written, has a numerator and denominator with a greatest common factor of 5, so it can be simplified to $\frac{7}{20}$.

Converting Percentages to Decimals: Because a percentage is based on "per hundred," decimals and percentages can be converted by multiplying or dividing by 100. Practically speaking, this always amounts to moving the decimal point two places to the right or left, depending on the conversion. To convert a percentage to a decimal, the decimal point is moved two places to the right and the % sign gets removed. To convert a decimal to a percentage, the decimal point is moved two places to the left and a "%" sign is added. Here are some examples:

65% = 0.65

0.33 = 33%

0.215 = 21.5%

99.99% = 0.9999

500% = 5.00

7.55 = 755%

Percentage Problems

Questions dealing with percentages can be difficult when they are phrased as word problems. These word problems almost always come in one of three varieties. The first type will ask to find what percentage of some number will equal another number. The second asks to determine what number is some percentage of another given number. The third will ask what number another number is a given percentage of.

One of the most important parts of correctly answering percentage word problems is to identify the numerator and the denominator. This fraction can then be converted into a percentage, as described in the previous section.

The following word problem shows how to make this conversion:

A department store carries several different types of footwear. The store is currently selling 8 athletic shoes, 7 dress shoes, and 5 sandals. What percentage of the store's footwear are sandals?

The first step is to calculate what serves as the 'whole', as this will be the denominator. How many total pieces of footwear does the store sell? The store sells 20 different types (8 athletic + 7 dress + 5 sandals).

In the next step, test takers need to determine which footwear type the question is specifically asking about: sandals. Thus, 5 is the numerator.

Lastly, the resultant fraction must be expressed as a percentage. The first two steps indicate that $\frac{5}{20}$ of the footwear pieces are sandals. This fraction must now be converted into a percentage: $\frac{5}{20} \times \frac{5}{5} = \frac{25}{100} = 25\%$.

Positive and Negative Numbers

Signs

Aside from 0, numbers can be either positive or negative. The sign for a positive number is the plus sign or the + symbol, while the sign for a negative number is minus sign or the – symbol. If a number has no designation, then it's assumed to be positive.

Absolute Values

Both positive and negative numbers are valued according to their distance from 0. Both +3 and -3 can be considered using the following number line:

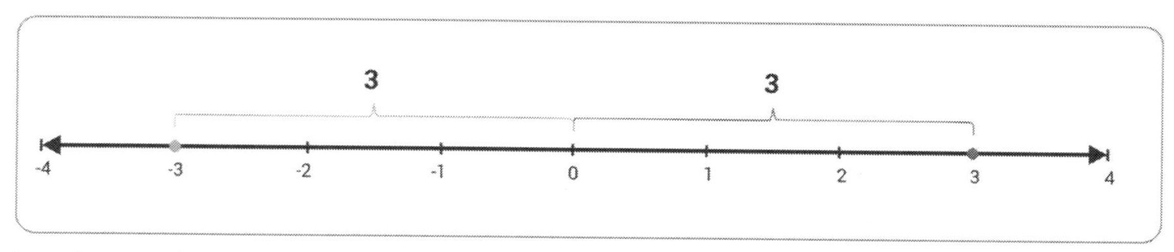

Both 3 and -3 are three spaces from 0. The distance from 0 is called its *absolute value*. Thus, both -3 and 3 have an absolute value of 3 since they're both three spaces away from 0.

An absolute number is written by placing | | around the number. So, $|3|$ and $|-3|$ both equal 3, as that's their common absolute value.

Algebra

Solving Equations with Exponents and Roots

Here are some of the most important properties of exponents and roots: if *n* is an integer, and if $a^n = b^n$, then $a = b$ if *n* is odd; but $a = \pm b$ if *n* is even. Similarly, if the roots of two numbers are equal, $\sqrt[n]{a} = \sqrt[n]{b}$, then $a = b$. This means that when starting with a true equation, both sides of that equation can be raised to a given power to obtain another true equation. Beware that when an even-powered root is taken on both sides of the equation, a $\pm$ is needed in the result. For example, given the equation $x^2 = 16$, one needs to take the square root of both sides to solve for x. This results in the answer $x = \pm 4$ because $(-4)^2 = 16$ and $(4)^2 = 16$.

Another property is that if $a^n = a^m$, then $n = m$. This is true for any real numbers *n* and *m*.

For solving the equation $\sqrt{x+2} - 1 = 3$, the first step is to move the -1 over to the right-hand side of the equation. This is performed by adding 1 to both sides, which yields $\sqrt{x+2} = 4$. Next, both sides are squared, remembering that by squaring both sides, the signs are irrelevant. This yields $x + 2 = 16$, which simplifies to give $x = 14$.

Now consider the problem $(x+1)^4 = 16$. To solve this, one takes the 4th root of both sides, which means an ambiguity in the sign will be introduced because it is an even root: $\sqrt[4]{(x+1)^4} = \pm\sqrt[4]{16}$. The right-hand side is 2, since $2^4 = 16$. Therefore, $x + 1 = \pm 2$, or $x = -1 \pm 2$. Thus, the two possible solutions are $x = -3$ and $x = 1$.

A helpful tip for test takers to remember is that when solving equations, the answer can be checked by plugging the solution back into the problem to make a true statement.

Relations and Functions

First, it's important to understand the definition of a *relation*. Given two variables, x and *y*, which stand for unknown numbers, a *relation* between x and *y* is an object that splits all of the pairs (x, *y*) into those for which the relation is true and those for which it is false. For example, consider the relation of $x^2 = y^2$. This relationship is true for the pair (1, 1) and for the pair (-2, 2), but false for (2, 3). Another example of a relation is $x \leq y$. This is true whenever x is less than or equal to *y*.

A *function* is a special kind of relation where, for each value of x, there is only a single value of *y* that satisfies the relation. So, $x^2 = y^2$ is *not* a function because in this case, if x is 1, *y* can be either 1 or -1: the pair (1, 1) and (1, -1) both satisfy the relation. More generally, for this relation, any pair of the form $(a, \pm a)$ will satisfy the relation. On the other hand, consider the following relation: $y = x^2 + 1$. This is a function because for each value of x, there is a unique value of *y* that satisfies the relation. Notice, however, there are multiple values of x that give us the same value of *y*. This is perfectly acceptable for a function. Therefore, *y* is a function of x.

To determine if a relation is a function, one should check to see if every x-value has a unique corresponding *y*-value.

A function can be viewed as an object that has x as its input and outputs a unique *y*-value. It is sometimes convenient to express this using *function notation*, where the function itself is given a name, often *f*. To emphasize that *f* takes x as its input, the function is written as $f(x)$. In the above example, the equation could be rewritten as $f(x) = x^2 + 1$. To write the value that a function yields for some specific value of x, that value is put in place of x in the function notation. For example, $f(3)$ will denote the value that the function outputs when the input value is 3. If $f(x) = x^2 + 1$, then $f(3) = 3^2 + 1 = 10$.

A function can also be viewed as a table of pairs (x, *y*), which lists the value for *y* for each possible value of x.

The set of all possible values for x in $f(x)$ is called the *domain* of the function, and the set of all possible outputs is called the *range* of the function. Note that usually the domain is assumed to be all real numbers, except those for which the expression for $f(x)$ is not defined, unless the problem specifies otherwise. An example of how a function might not be defined is in the case of $f(x) = \frac{1}{x+1}$, which is not defined when $x = -1$ (which would require dividing by zero). Therefore, in this case, the domain would be all real numbers except $x = -1$.

If *y* is a function of x, then x is the *independent variable* and *y* is the *dependent variable*. This is because in many cases, the problem will start with some value of x and then see how *y* changes depending on this starting value.

Evaluating Functions

To evaluate functions, the given value is plugged in everywhere that the variable appears in the expression for the function. For example, find $g(-2)$ where $g(x) = 2x^2 - \frac{4}{x}$. To complete the problem, -2 is plugged in in the following way: $g(-2) = 2(-2)^2 - \frac{4}{-2} = 2 \cdot 4 + 2 = 8 + 2 = 10$.

Defining Linear Equations

A function is considered *linear* if it can take the form of the equation $f(x) = ax + b$, or $y = ax + b$, for any two numbers *a* and *b*. A linear equation forms a straight line when graphed on the coordinate plane. An example of a linear function is shown below on the graph.

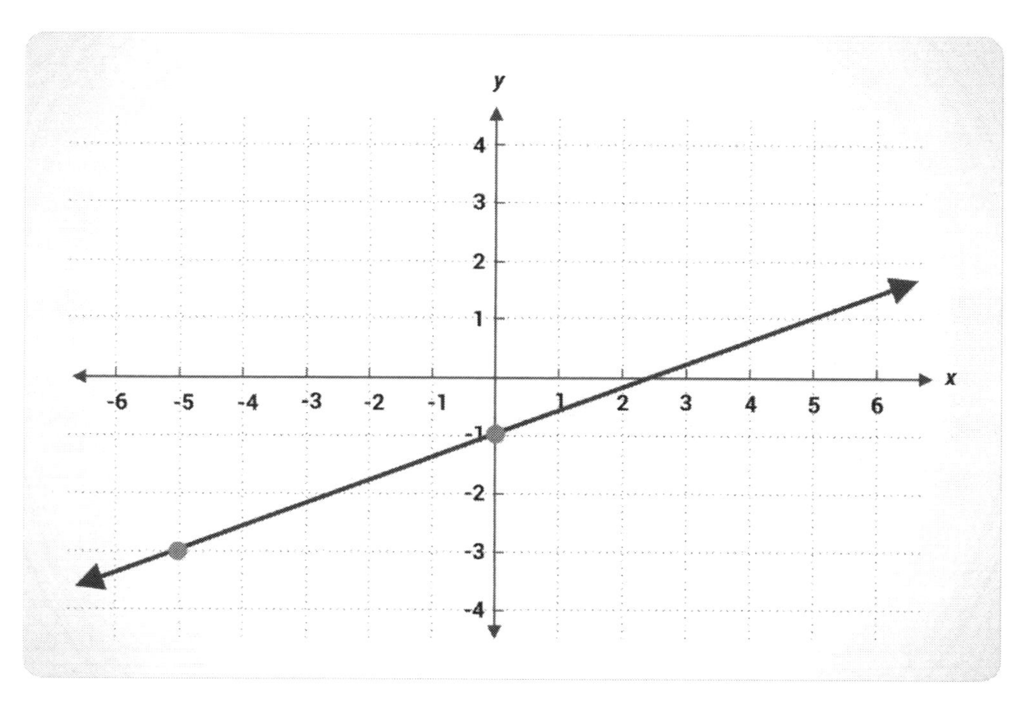

This is a graph of the function $y = \frac{2}{5}x - 1$. A table of values that satisfies this function is shown below.

x	y
-5	-3
0	-1
5	1
10	3

These points can be found on the graph using the form (x,y). For more on graphing in the coordinate plane, refer to the *Graphing* section below.

Graphing Functions and Relations

To graph relations and functions, the Cartesian plane is used. This can be visualized as a plane with a grid of squares, with one direction being the x-axis and the other direction the *y*-axis. Generally, the independent variable is placed along the horizontal axis, and the dependent variable is placed along the vertical axis. Any point on the plane can be specified by identifying the point's location along each of the two axes with a pair of numbers (x, y). Specific values for these pairs can be given names such as $C = (-1, 3)$. Negative values mean to move left or down; positive values mean to move right or up. The point where the axes cross one another is called the *origin*. The origin has coordinates $(0, 0)$ and is usually called *O* when given a specific label.

An illustration of the Cartesian plane, along with graphs of $(2, 1)$ and $(-1, -1)$, are below:

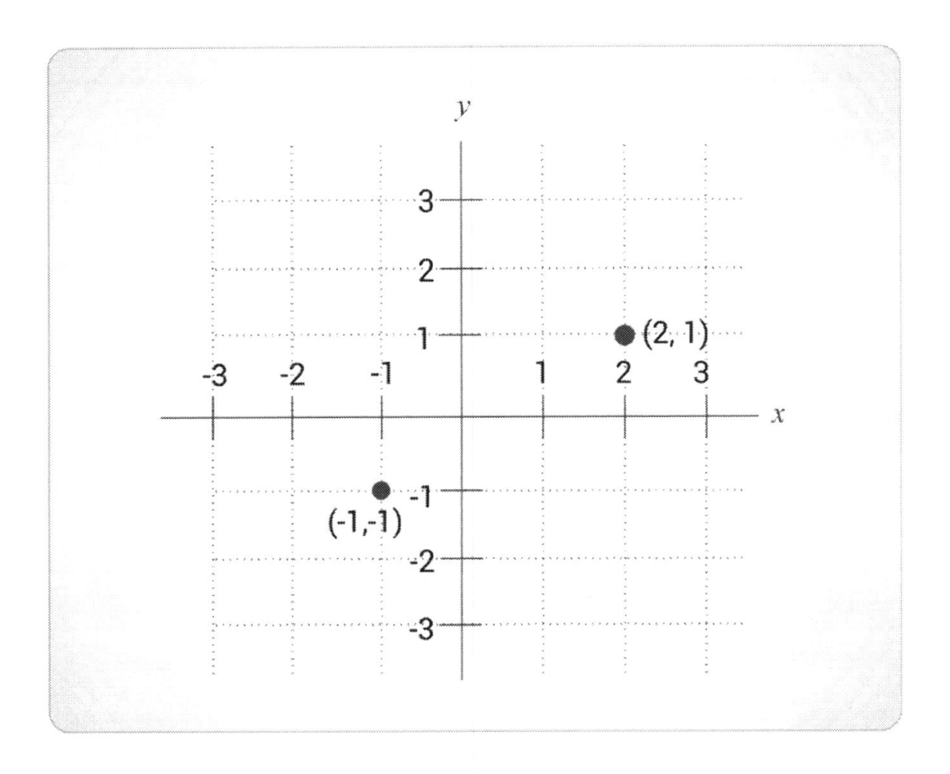

Relations also can be graphed by marking each point whose coordinates satisfy the relation. If the relation is a function, then there is only one value of y for any given value of x. This leads to the *vertical line test*: if a relation is graphed, then the relation is a function if every vertical line touches the graph at either zero or one point. Conversely, when graphing a function, then every vertical line will touch the graph at no points or just one point.

Forms of Linear Equations

When graphing a linear function, the ratio of the change of the y-coordinate to the change in the x-coordinate is constant between any two points on the resulting line, no matter which two points are chosen. In other words, in a pair of points on a line, (x_1, y_1) and (x_2, y_2), with $x_1 \neq x_2$ so that the two points are distinct, then the ratio $\frac{y_2 - y_1}{x_2 - x_1}$ will be the same, regardless of which particular pair of points are chosen. This ratio, $\frac{y_2 - y_1}{x_2 - x_1}$, is called the *slope* of the line and is frequently denoted with the letter m. If slope *m* is positive, then the line goes upward when moving to the right, while if slope *m* is negative, then the line goes downward when moving to the right. If the slope is 0, then the line is called *horizontal*, and the y-coordinate is constant along the entire line. In lines where the x-coordinate is constant along the entire line, y is not actually a function of x. For such lines, the slope is not defined. These lines are called *vertical* lines.

Linear functions may take forms other than $y = ax + b$. The most common forms of linear equations are explained below:

- Standard Form: $Ax + By = C$, in which the slope is given by $m = \frac{-A}{B}$, and the y-intercept is given by $\frac{C}{B}$.
- Slope-Intercept Form: $y = mx + b$, where the slope is *m* and the *y*-intercept is *b*.
- Point-Slope Form: $y - y_1 = m(x - x_1)$, where the slope is *m* and (x_1, y_1) is any point on the chosen line.
- Two-Point Form: $\frac{y-y_1}{x-x_1} = \frac{y_2-y_1}{x_2-x_1}$, where (x_1, y_1) and (x_2, y_2) are any two distinct points on the chosen line. Note that the slope is given by $m = \frac{y_2-y_1}{x_2-x_1}$.
- Intercept Form: $\frac{x}{x_1} + \frac{y}{y_1} = 1$, in which x_1 is the x-intercept and y_1 is the y-intercept.

These five ways to write linear equations are all useful in different circumstances. Depending on the given information, it may be easier to write one of the forms over another.

If $y = mx$, *y* is directly proportional to x. In this case, changing x by a factor changes *y* by that same factor. If $y = \frac{m}{x}$, *y* is inversely proportional to x. For example, if x is increased by a factor of 3, then *y* will be decreased by the same factor, 3.

Solving Linear Equations

Sometimes, rather than a situation where there's an equation such as $y = ax + b$ and the goal is to find *y* for some value of x, the result is given and finding x is requested.

The key to solving any equation is to remember that from one true equation, another true equation can be found by adding, subtracting, multiplying, or dividing both sides by the same quantity. In this case, it's necessary to manipulate the equation so that one side only contains x. Then the other side will show what x is equal to.

For example, in solving $3x - 5 = 2$, adding 5 to each side results in $3x = 7$. Next, dividing both sides by 3 results in $x = \frac{7}{3}$. To ensure the value of x is correct, the number can be substituted into the original equation and solved to see if it makes a true statement. For example, $3(\frac{7}{3}) - 5 = 2$ can be simplified by cancelling out the two 3s. This yields $7 - 5 = 2$, which is a true statement.

Sometimes an equation may have more than one x term. For example, consider the following equation: $3x + 2 = x - 4$. Moving all of the x terms to one side by subtracting x from both sides results in $2x + 2 = -4$. Next, 2 is subtracted from both sides so that there is no constant term on the left side. This yields $2x = -6$. Finally, both sides are divided by 2, which leaves $x = -3$.

Solving Linear Inequalities

Solving linear inequalities is very similar to solving equations, except for one rule: when multiplying or dividing an inequality by a negative number, the inequality symbol changes direction. Given the following inequality, solve for x: $-2x + 5 < 13$. The first step in solving this equation is to subtract 5 from both sides. This leaves the inequality: $-2x < 8$. The last step is to divide both sides by -2. By using the rule, the answer to the inequality is $x > -4$.

Since solutions to inequalities include more than one value, number lines are used many times to model the answer. For the previous example, the answer is modelled on the number line below. It shows that any number greater than -4, not including -4, satisfies the inequality.

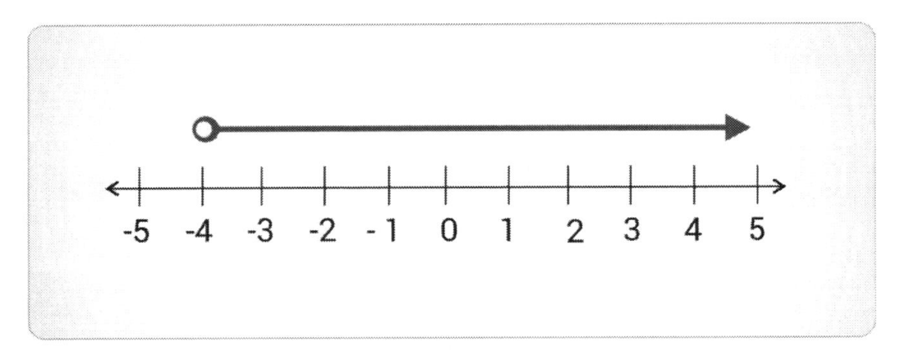

Linear Systems of Equations

A problem sometimes involves multiple variables and multiple equations. These are called *systems of equations*. In this case, one should try to manipulate them until an expression is found that provides the value of one of the variables. There are a couple of different approaches to this, and in some cases, some of them can be used together. The three basic rules to keep in mind are the following:

- A set of equations can be manipulated by performing the same operation to both equations, just as is done when working with just one equation.
- If one of the equations can be changed so that it expresses one variable in terms of the others, then that expression can be substituted into the other equations and the variable can be eliminated. This means the other equations will have one less variable in them. This is called the *method of substitution*.
- If two equations of the form $a = b, c = d$ are included, then a new equation can be formed by adding the left sides and adding the right sides, $a + c = b + d$, or $a - c = b - d$. This enables the elimination of one of the variables from an equation. This is called the *method of elimination.*

The simplest case is the case of a *linear* system of equations. Although the equations may be written in more complicated forms, linear systems of equations with two variables can always be written in the form $ax + by = c, dx + ey = f$. The two basic approaches to solving these systems are substitution and elimination.

Consider the system $3x - y = 2$ and $2x + 2y = 3$. This can be solved in two ways.

- By substitution: start by solving the first equation for y. First, subtract $3x$ from both sides to obtain $-y = 2 - 3x$. Next, divide both sides by -1, to obtain $y = 3x - 2$. Then substitute this value for y into the second equation. This yields $2x + 2(3x - 2) = 3$. This can be simplified to $2x + 6x - 4 = 3$, or $8x = 7$, which means $x = \frac{7}{8}$. Plugging in this value for x into $y = 3x - 2$, yields $y = 3\left(\frac{7}{8}\right) - 2 = \frac{21}{8} - \frac{16}{8} = \frac{5}{8}$. So, this results in $x = \frac{7}{8}, y = \frac{5}{8}$.
- By elimination: first, multiply the first equation by 2. This results in $-2y$, which could cancel out the $+2y$ in the second equation. Multiplying both sides of the first equation by 2 gives results in $2(3x - y) = 2(2)$, or $6x - 2y = 4$. Adding the left sides and the right sides of the two

equations and setting the results equal to one another results in $(6x + 2x) + (-2y + 2y) = 4 + 3$. This simplifies to $8x = 7$, so again $x = \frac{7}{8}$. Plug this back into either of the original equations and the result is $3\left(\frac{7}{8}\right) - y = 2$ or $y = 3\left(\frac{7}{8}\right) - 2 = \frac{21}{8} - \frac{16}{8} = \frac{5}{8}$. This again yields $x = \frac{7}{8}, y = \frac{5}{8}$.

As this shows, both methods will give the same answer. However, one method is sometimes preferred over another simply because of the amount of work required. To check the answer, the values can be substituted into the given system to make sure they form two true statements.

Quadratic Functions

A polynomial of degree 2 is called *quadratic*. Every quadratic function can be written in the form $ax^2 + bx + c$. The graph of a quadratic function, $y = ax^2 + bx + c$, is called a *parabola*. Parabolas are vaguely U-shaped.

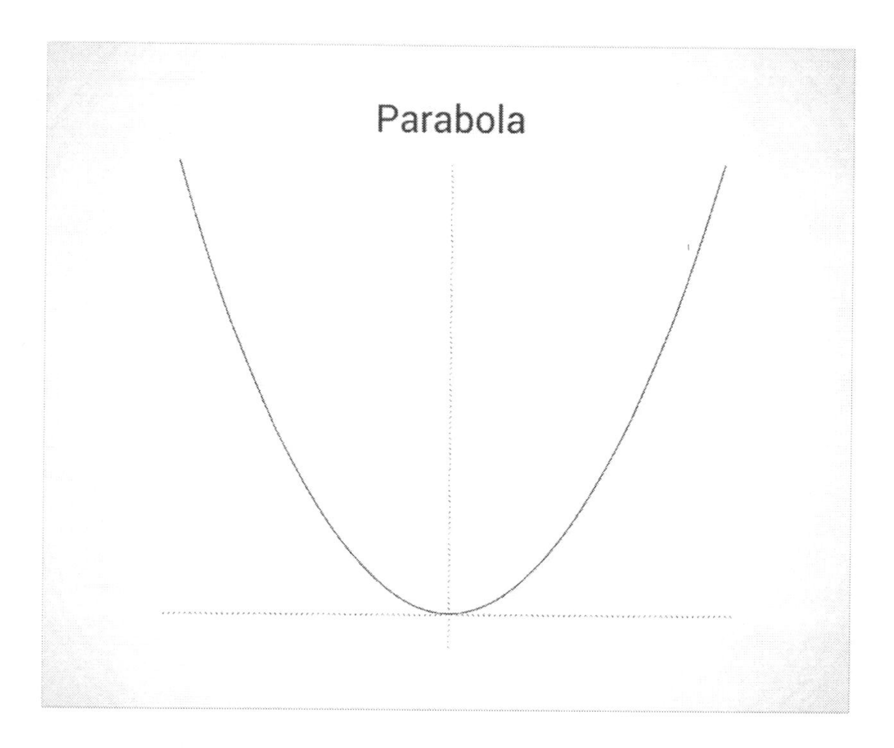

Whether the parabola opens upward or downward depends on the sign of *a*. If *a* is positive, then the parabola will open upward. If *a* is negative, then the parabola will open downward. The value of *a* will also affect how wide the parabola is. If the absolute value of *a* is large, then the parabola will be fairly skinny. If the absolute value of *a* is small, then the parabola will be quite wide.

Changes to the value of *b* affect the parabola in different ways, depending on the sign of *a*. For positive values of *a*, increasing *b* will move the parabola to the left, and decreasing *b* will move the parabola to the right. On the other hand, if *a* is negative, the effects will be the opposite: increasing *b* will move the parabola to the right, while decreasing *b* will move the parabola to the left.

Changes to the value of *c* move the parabola vertically. The larger that *c* is, the higher the parabola gets. This does not depend on the value of *a*.

The quantity $D = b^2 - 4ac$ is called the *discriminant* of the parabola. When the discriminant is positive, then the parabola has two real zeros, or x-intercepts. However, if the discriminant is negative, then there are no real zeros, and the parabola will not cross the x-axis. The highest or lowest point of the parabola is called the *vertex*. If the discriminant is zero, then the parabola's highest or lowest point is on the x-axis, and it will have a single real zero. The x-coordinate of the vertex can be found using the equation $x = -\frac{b}{2a}$. This x-value can be plugged into the equation to find the y-coordinate.

A quadratic equation is often used to model the path of an object thrown into the air. The x-value can represent the time in the air, while the y-value can represent the height of the object. In this case, the maximum height of the object would be the y-value found when the x-value is $-\frac{b}{2a}$.

Solving Quadratic Equations

A *quadratic equation* is an equation in the form $ax^2 + bx + c = 0$. There are several methods to solve such equations. The easiest method will depend on the particular quadratic equation in question.

It is sometimes possible to solve quadratic equations by manually *factoring* them. This means rewriting them in the form $(x + A)(x + B) = 0$. If this is done, then they can be solved by remembering that when $ab = 0$, either *a* or *b* must be equal to zero. Therefore, to have $(x + A)(x + B) = 0$, $(x + A) = 0$ or $(x + B) = 0$ is needed. These equations have the solutions $x = -A$ and $x = -B$, respectively.

In order to factor a quadratic equation, note that $(x + A)(x + B) = x^2 + (A + B)x + AB$. So, if an equation is in the form $x^2 + bx + c$, two numbers, *A* and *B,* need to be found that will add up to *b*, and multiply together to give *c*.

As an example, consider solving the equation $-3x^2 + 6x + 9 = 0$. The first step is to divide both sides by -3, which yields $x^2 - 2x - 3 = 0$. Because $1 - 3 = -2$ and $(1)(-3) = -3$, the equation can be factored into $(x + 1)(x - 3) = 0$. Now, one can solve $(x + 1) = 0$ and $(x - 3) = 0$ to get the solutions $x = -1$ and $x = 3$.

When trying to factor, it is useful to remember that $x^2 + 2xy + y^2 = (x + y)^2$, $x^2 - 2xy + y^2 = (x - y)^2$, and $x^2 - y^2 = (x + y)(x - y)$.

However, factoring by hand is often hard to do. If there are no obvious ways to factor the quadratic equation, solutions can still be found by using the *quadratic formula*.

The quadratic formula is $x = \frac{-b \pm \sqrt{b^2 - 4ac}}{2a}$. This method will always work, although it sometimes can take longer than factoring by hand, which can be quick if the factors are easy to guess. Using the standard form $ax^2 + bx + c = 0$, the values of *a*, *b*, and *c* from the equation can be plugged into the formula to solve for x. There will either be two answers, one answer, or no real answer. No real answer comes when the value of the *discriminant*—the number under the square root—is a negative number. Since there are no real numbers that, when squared, result in a negative, the answer will be no real roots.

Here is an example of solving a quadratic equation using the quadratic formula. Suppose the equation to solve is $-2x^2 + 3x + 1 = 0$. There is no obvious way to factor this, so the quadratic formula is used, with $a = -2, b = 3, c = 1$. After substituting these values into the quadratic formula, it yields:

$$x = \frac{-3 \pm \sqrt{3^2 - 4(-2)(1)}}{2(-2)}$$

This can be simplified to obtain:

$$\frac{3 \pm \sqrt{9+8}}{4} \text{ or } \frac{3 \pm \sqrt{9+8}}{4}.$$

Challenges can be encountered when asked to find a quadratic equation with specific roots. Given roots *A* and *B*, a quadratic function can be constructed with those roots by taking $(x - A)(x - B)$. So, constructing a quadratic equation with roots $x = -2, 3$, would result in $(x + 2)(x - 3) = x^2 - x - 6$. Multiplying this by a constant also could be done without changing the roots.

Geometry and Measurement

Lines, Rays, and Line Segments

The basic unit of geometry is a point. A *point* represents an exact location on a plane, or flat surface. The position of a point is indicated with a dot and usually named with a single uppercase letter, such as point *A* or point *T*. A point is a place, not a thing, and therefore has no dimensions or size. A set of points that lie on the same line are considered *collinear*. A set of points that lie in the same plane are *coplanar*.

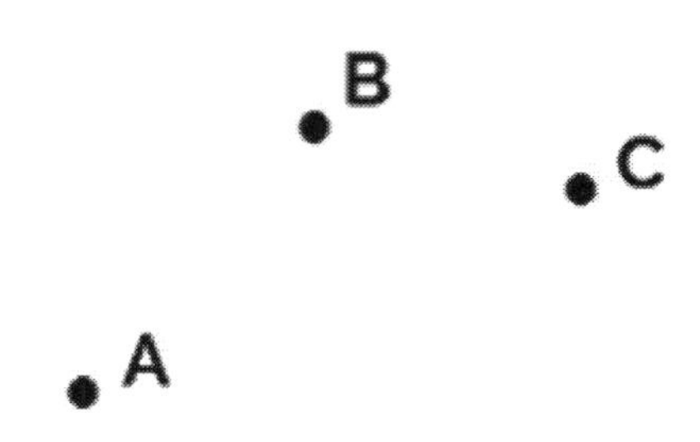

The image above displays point *A*, point *B*, and point *C*.

A *line* is as series of points that extends in both directions without ending. It consists of an infinite number of points and is drawn with arrows on both ends to indicate it that extends infinitely. Lines can be named by two points on the line or with a single, cursive, lowercase letter.

Two lines are considered *parallel* to each other if, while extending infinitely, they will never intersect (or meet). Parallel lines point in the same direction and are always the same distance apart. Two lines are considered *perpendicular* if they form a right angle at their intersection. Right angles are 90°. Typically, a small box is drawn at the intersection point to indicate the right angle.

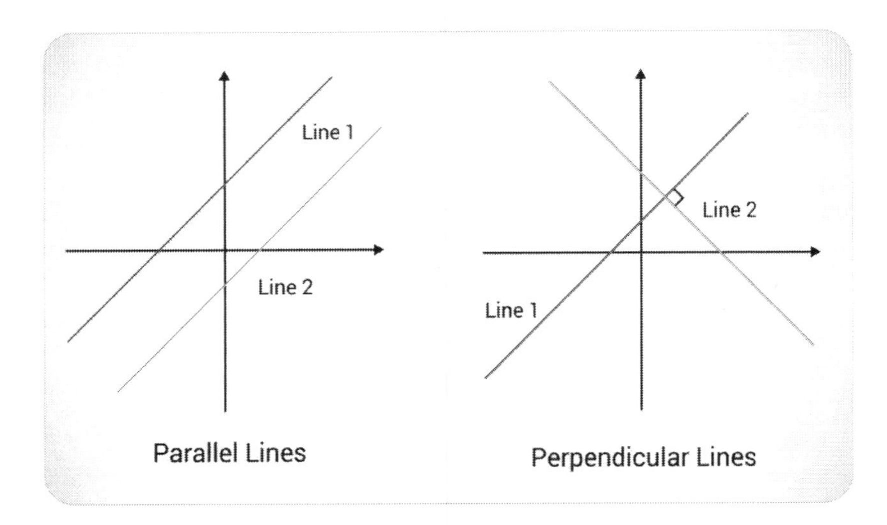

Line 1 is parallel to line 2 in the left image and is written as line 1 || line 2. Line 1 is perpendicular to line 2 in the right image and is written as line 1 $\perp$ line 2.

A *ray* has a specific starting point and extends in one direction without ending. The endpoint of a ray is its starting point. Rays are named using the endpoint first, and any other point on the ray. The following ray can be named ray *AB* and written $\overrightarrow{AB}$.

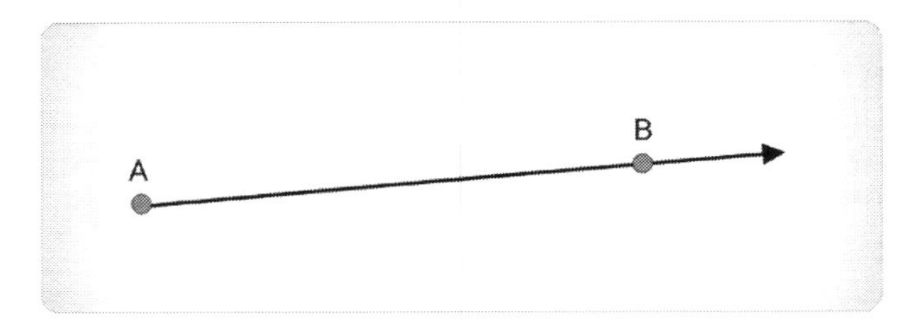

A *line segment* has specific starting and ending points. A line segment consists of two endpoints and all the points in between. Line segments are named by the two endpoints. The example below is named segment *KL* or segment *LK*, written $\overline{KL}$ or $\overline{LK}$.

Polygons and Solids

A *polygon* is a closed two-dimensional figure consisting of three or more sides. Polygons can be either convex or concave. A polygon that has interior angles all measuring less than 180° is *convex*. A *concave* polygon has one or more interior angles measuring greater than 180°.

Polygons can be classified by the number of sides (also equal to the number of angles) they have. The following are the names of polygons with a given number of sides or angles:

# of sides	3	4	5	6	7	8	9	10
Name of polygon	Triangle	Quadrilateral	Pentagon	Hexagon	Septagon (or heptagon)	Octagon	Nonagon	Decagon

Equiangular polygons are polygons in which the measure of every interior angle is the same. The sides of equilateral polygons are always the same length. If a polygon is both equiangular and equilateral, the polygon is defined as a *regular polygon*. Examples are shown below.

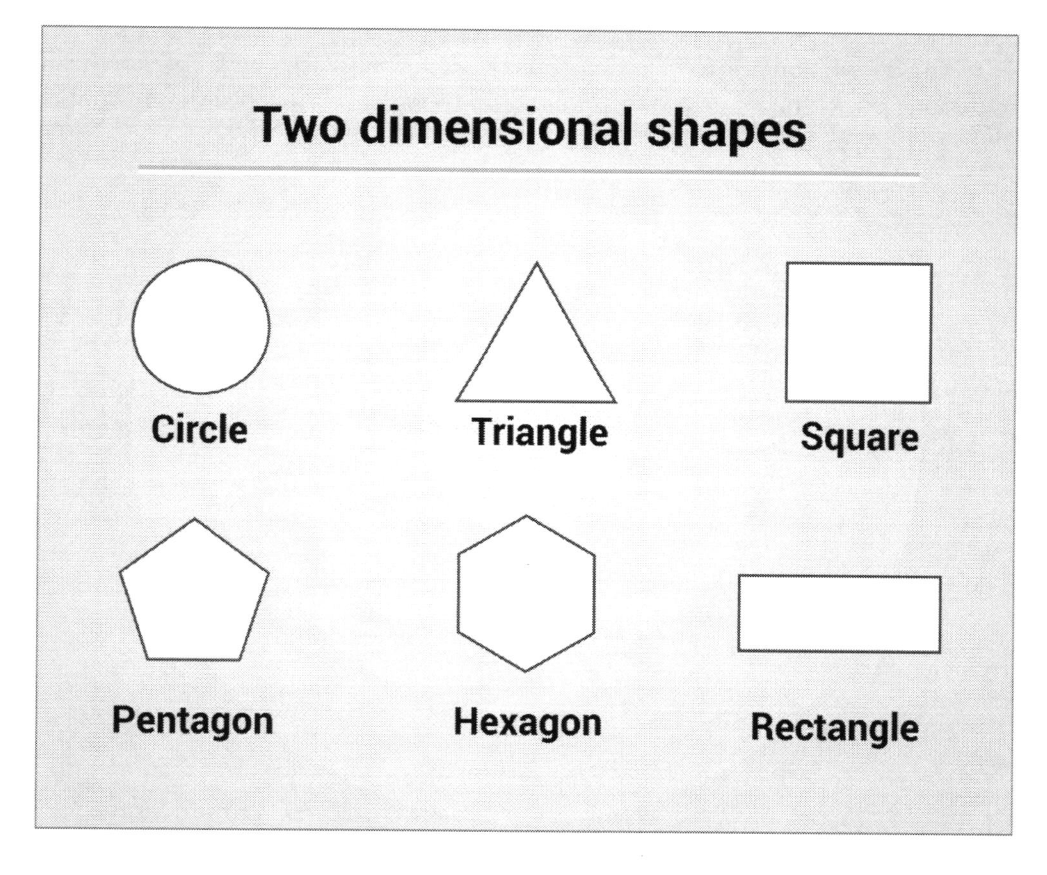

Triangles can be further classified by their sides and angles. A triangle with its largest angle measuring 90° is a *right triangle*. A triangle with the largest angle less than 90° is an *acute triangle*. A triangle with the largest angle greater than 90° is an *obtuse triangle*. Below is an example of a right triangle.

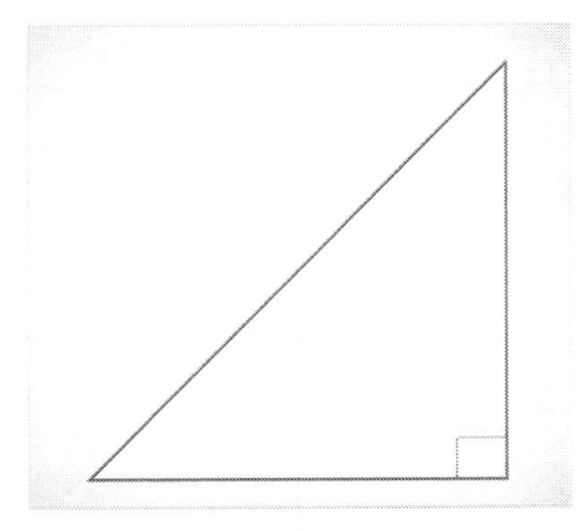

A triangle consisting of two equal sides and two equal angles is an *isosceles triangle*. A triangle with three equal sides and three equal angles is an *equilateral triangle*. A triangle with no equal sides or angles is a *scalene triangle*.

Isosceles triangle:

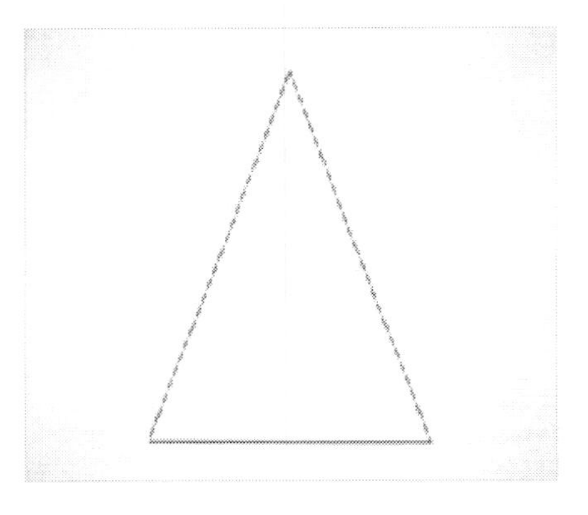

Equilateral triangle:

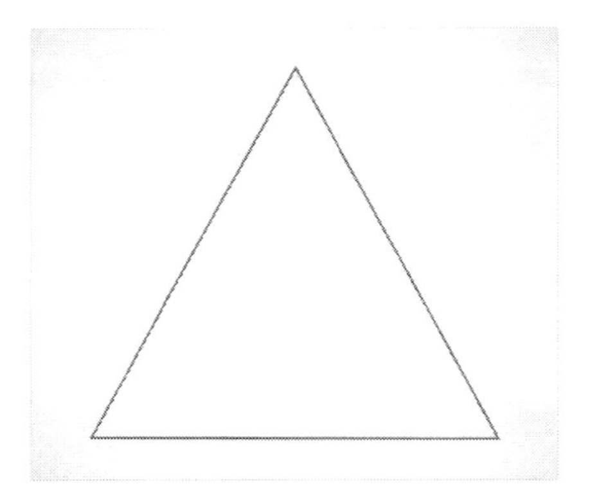

Scalene triangle:

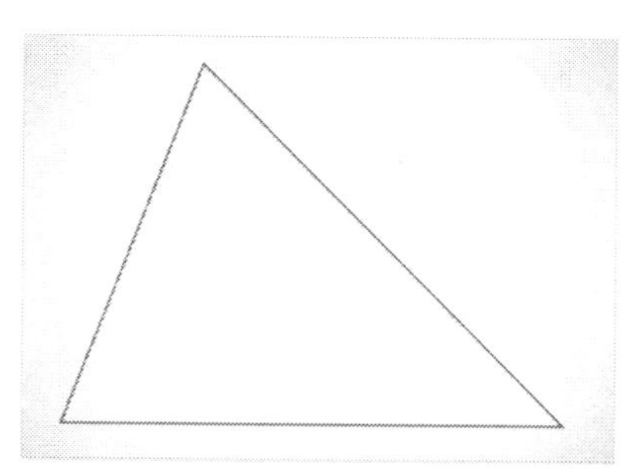

Quadrilaterals can be further classified according to their sides and angles. A quadrilateral with exactly one pair of parallel sides is called a *trapezoid*. A quadrilateral that shows both pairs of opposite sides parallel is a *parallelogram*. Parallelograms include rhombuses, rectangles, and squares. A *rhombus* has four equal sides. A rectangle has four equal angles (90° each). A square has four 90° angles and four equal sides. Therefore, a square is both a rhombus and a rectangle.

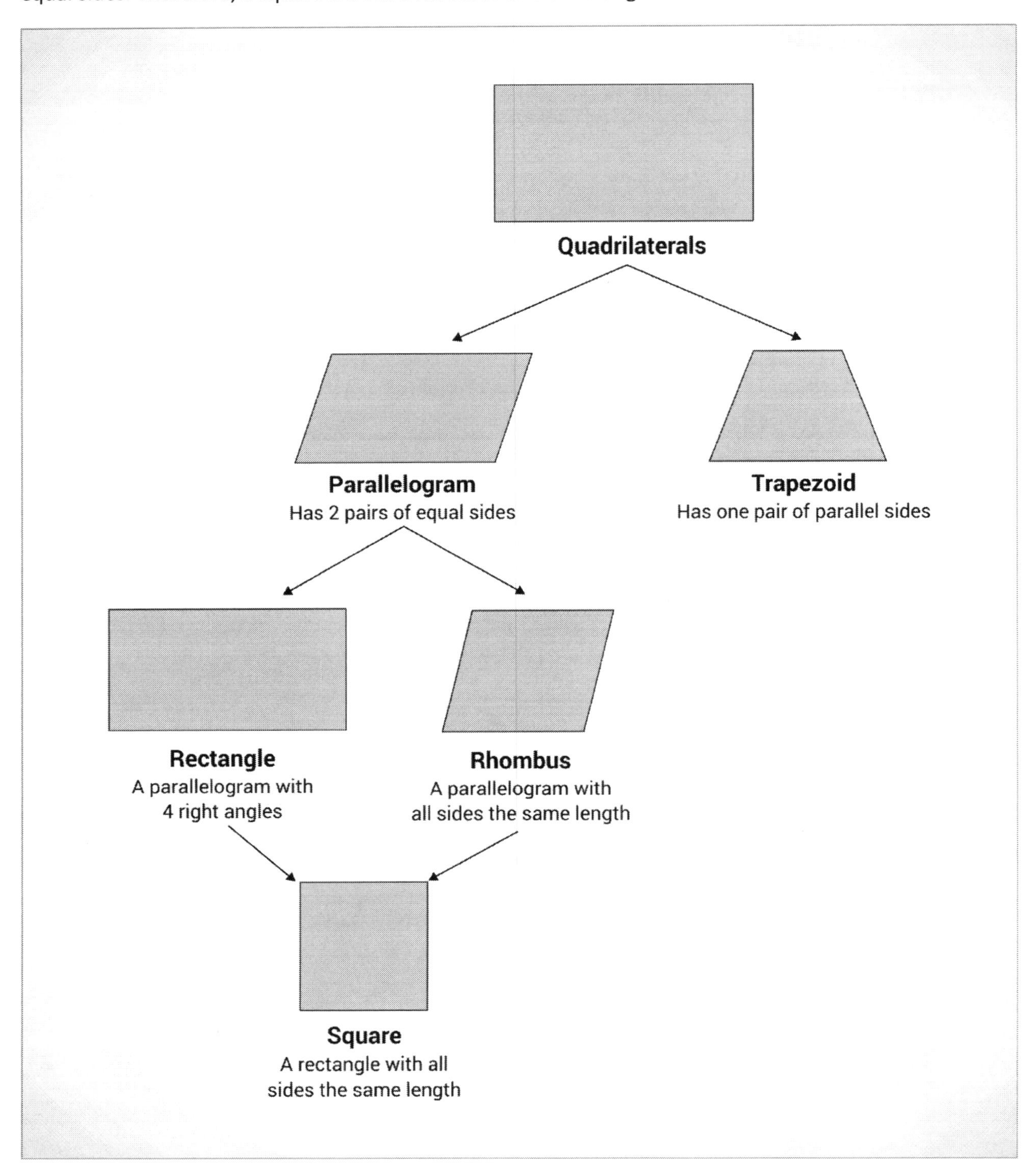

<u>Three-Dimensional Figures with Nets</u>

A net is a construction of two-dimensional figures that can be folded to form a given three-dimensional figure. More than one net may exist that will fold to produce the same solid, or three-dimensional figure. The bases and faces of the solid figure are analyzed to determine the polygons (two-dimensional figures) needed to form the net.

Consider the following triangular prism:

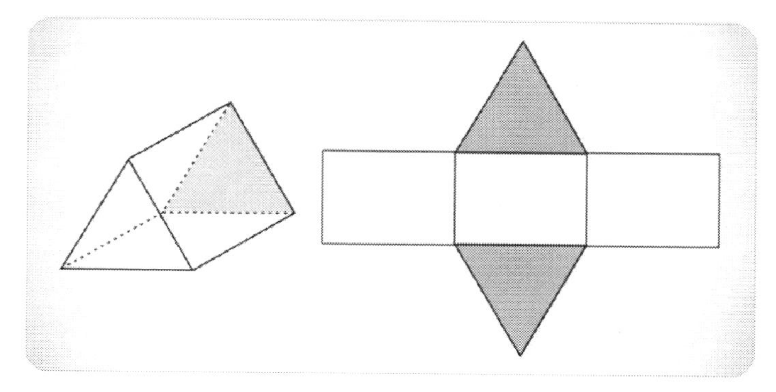

The surface of the prism consists of two triangular bases and three rectangular faces. The net beside it can be used to construct the triangular prism by first folding the triangles up to be parallel to each other, and then folding the two outside rectangles up and to the center with the outer edges touching.

Consider the following cylinder:

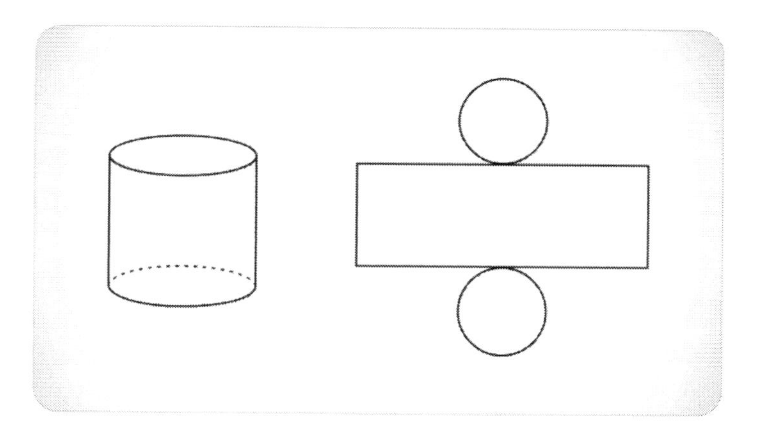

The surface consists of two circular bases and a curved lateral surface that can be opened and flattened into a rectangle. The net beside the cylinder can be used to construct the cylinder by first folding the circles up to be parallel to each other, and then curving the sides of the rectangle up to touch each other. The top and bottom of the folded rectangle should be touching the outside of both circles.

Consider the following square pyramid below on the left. The surface consists of one square base and four triangular faces. The net below on the right can be used to construct the square pyramid by folding each triangle towards the center of the square. The top points of the triangle meet at the vertex.

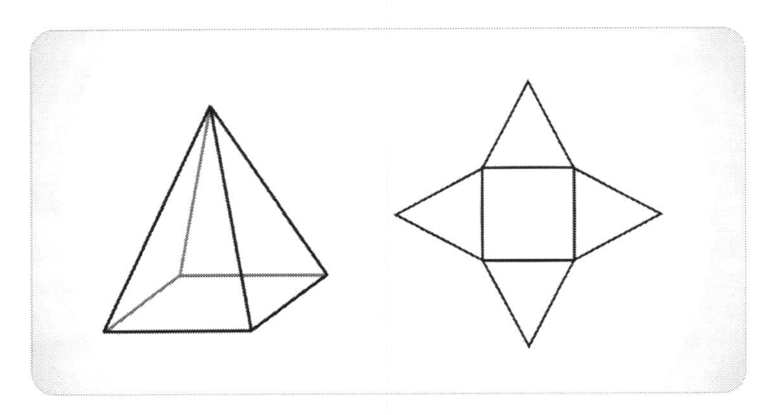

Surface Area of Three-Dimensional Figures

The *area* of a two-dimensional figure refers to the number of square units needed to cover the interior region of the figure. This concept is similar to wallpaper covering the flat surface of a wall. For example, if a rectangle has an area of 10 square centimeters (written 10cm^2), it will take 10 squares, each with sides one centimeter in length, to cover the interior region of the rectangle. Note that area is measured in square units such as: square centimeters or cm^2; square feet or ft^2; square yards or yd^2; square miles or mi^2.

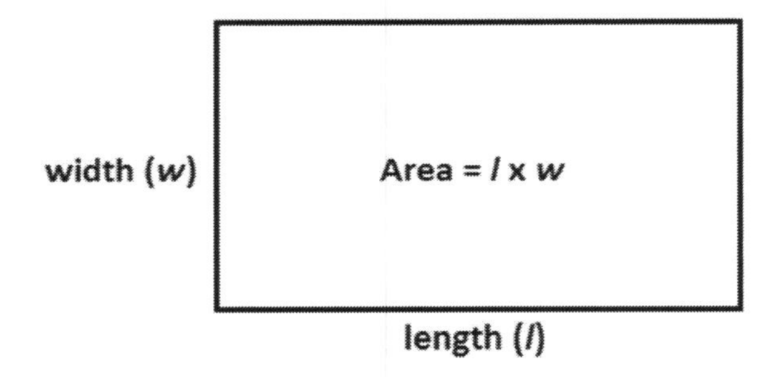

The *surface area* of a three-dimensional figure refers to the number of square units needed to cover the entire surface of the figure. This concept is similar to using wrapping paper to completely cover the outside of a box. For example, if a triangular pyramid has a surface area of 17 square inches (written 17in^2), it will take 17 squares, each with sides one inch in length, to cover the entire surface of the pyramid. Surface area is also measured in square units.

Many three-dimensional figures (solid figures) can be represented by nets consisting of rectangles and triangles. The surface area of such solids can be determined by adding the areas of each of its faces and bases. Finding the surface area using this method requires calculating the areas of rectangles and triangles. To find the area (*A*) of a rectangle, the length (*l*) is multiplied by the width $(w) \rightarrow A = l \times w$. The area of the rectangle below is calculated: $A = (8cm) \times (4cm) \rightarrow A = 32cm^2$.

To calculate the area (*A*) of a triangle, the product of $\frac{1}{2}$, the base (*b*), and the height (*h*) is found $\rightarrow A = \frac{1}{2} \times b \times h$. Note that the height of a triangle is measured from the base to the vertex opposite of it forming a right angle with the base. The area of the triangle below is calculated: $A = \frac{1}{2} \times (11cm) \times (6cm) \rightarrow A = 33cm^2$.

Consider the following triangular prism, which is represented by a net consisting of two triangles and three rectangles:

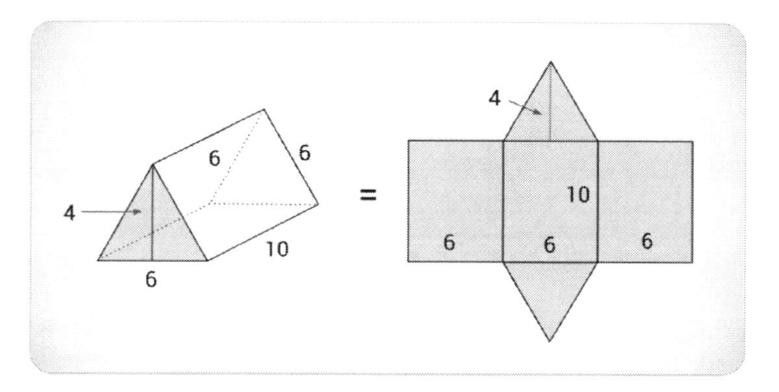

The surface area of the prism can be determined by adding the areas of each of its faces and bases. The surface area (*SA*) = area of triangle + area of triangle + area of rectangle + area of rectangle + area of rectangle.

$$SA = \left(\frac{1}{2} \times b \times h\right) + \left(\frac{1}{2} \times b \times h\right) + (l \times w) + (l \times w) + (l \times w)$$

$$SA = \left(\frac{1}{2} \times 6 \times 4\right) + \left(\frac{1}{2} \times 6 \times 4\right) + (6 \times 10) + (6 \times 10) + (6 \times 10)$$

$$SA = (12) + (12) + (60) + (60) + (60)$$

$$SA = 204 \text{ square units}$$

Area and Perimeter of Polygons

Perimeter is the measurement of a distance around something or the sum of all sides of a polygon. Think of perimeter as the length of the boundary, like a fence. In contrast, *area* is the space occupied by a defined enclosure, like a field enclosed by a fence.

When thinking about perimeter, think about walking around the outside of something. When thinking about area, think about the amount of space or *surface area* something takes up.

Squares

The perimeter of a square is measured by adding together all of the sides. Since a square has four equal sides, its perimeter can be calculated by multiplying the length of one side by 4. Thus, the formula is $P = 4 \times s$, where s equals one side. For example, the following square has side lengths of 5 meters:

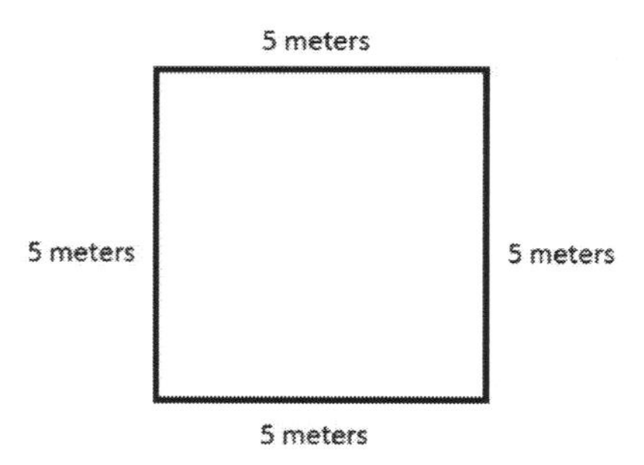

The perimeter is 20 meters because 4 times 5 is 20.

The area of a square is the length of a side squared, and the area of a rectangle is length multiplied by the width. For example, if the length of the square is 7 centimeters, then the area is 49 square centimeters. The formula for this example is $A = s^2 = 7^2 = 49$ square centimeters. An example is if the rectangle has a length of 6 inches and a width of 7 inches, then the area is 42 square inches:

$$A = lw = 6(7) = 42 \text{ square inches}$$

Rectangles

Like a square, a rectangle's perimeter is measured by adding together all of the sides. But as the sides are unequal, the formula is different. A rectangle has equal values for its lengths (long sides) and equal values for its widths (short sides), so the perimeter formula for a rectangle is:

$$P = l + l + w + w = 2l + 2w$$

l equals length
w equals width

The area is found by multiplying the length by the width, so the formula is $A = l \times w$.

For example, if the length of a rectangle is 10 inches and the width 8 inches, then the perimeter is 36 inches because:

$$P = 2l + 2w = 2(10) + 2(8) = 20 + 16 = 36 \text{ inches}$$

Triangles

A triangle's perimeter is measured by adding together the three sides, so the formula is $P = a + b + c$, where a, b, and c are the values of the three sides. The area is the product of one-half the base and height so the formula is:

$$A = \frac{1}{2} \times b \times h$$

It can be simplified to:

$$A = \frac{bh}{2}$$

The base is the bottom of the triangle, and the height is the distance from the base to the peak. If a problem asks to calculate the area of a triangle, it will provide the base and height.

For example, if the base of the triangle is 2 feet and the height 4 feet, then the area is 4 square feet. The following equation shows the formula used to calculate the area of the triangle:

$$\text{A} = \frac{1}{2}\text{bh} = \frac{1}{2}(2)(4) = 4 \text{ square feet}$$

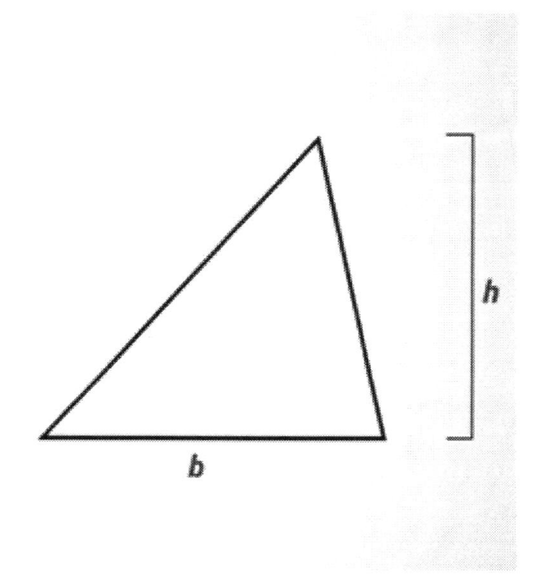

Parallelograms
Similar to triangles, the height of the parallelogram is measured from one base to the other at a 90° angle (or perpendicular).

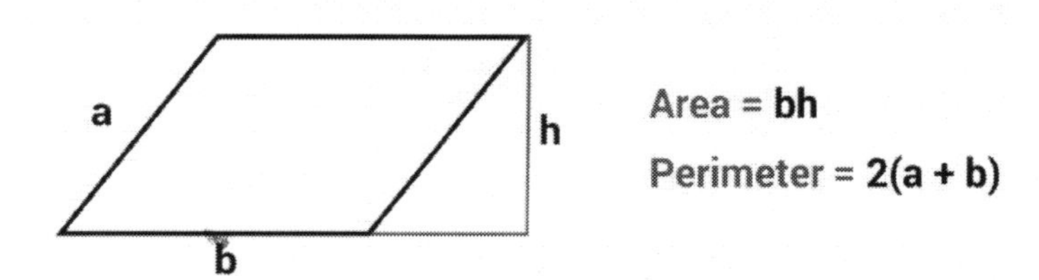

Trapezoid

The area of a trapezoid can be calculated using the formula: $A = \frac{1}{2} \times h(b_1 + b_2)$, where *h* is the height and b_1 and b_2 are the parallel bases of the trapezoid.

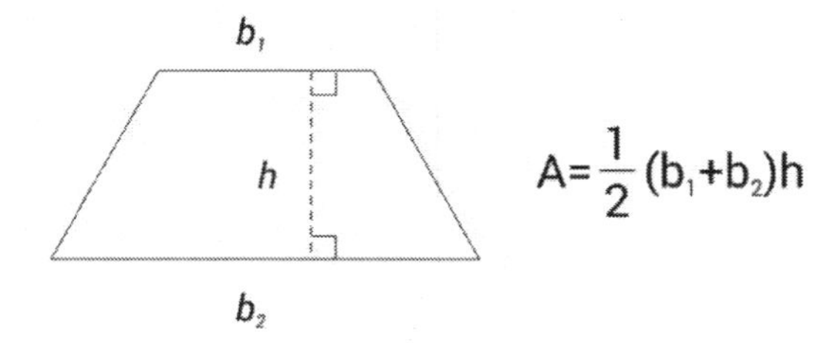

Regular Polygon

The area of a regular polygon can be determined by using its perimeter and the length of the apothem. The apothem is a line from the center of the regular polygon to any of its sides at a right angle. (Note that the perimeter of a regular polygon can be determined given the length of only one side.) The formula for the area (*A*) of a regular polygon is $A = \frac{1}{2} \times a \times P$, where *a* is the length of the apothem and *P* is the perimeter of the figure. Consider the following regular pentagon:

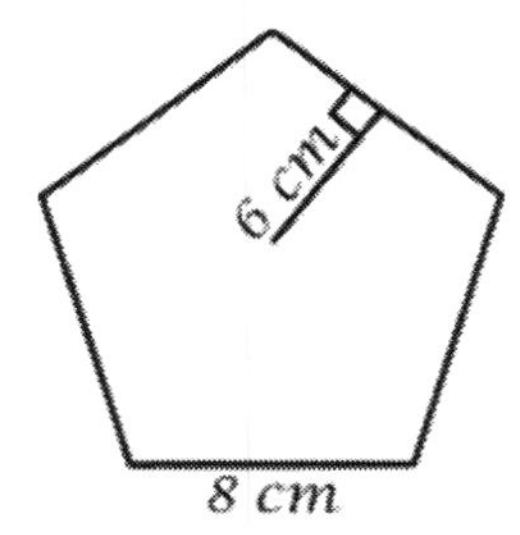

To find the area, the perimeter (*P*) is calculated first: $8cm \times 5 \rightarrow P = 40cm$. Then the perimeter and the apothem are used to find the area (*A*): $A = \frac{1}{2} \times a \times P \rightarrow A = \frac{1}{2} \times (6cm) \times (40cm) \rightarrow A = 120cm^2$. Note that the unit is $cm^2 \rightarrow cm \times cm = cm^2$.

Irregular Shapes

The perimeter of an irregular polygon is found by adding the lengths of all of the sides. In cases where all of the sides are given, this will be very straightforward, as it will simply involve finding the sum of the provided lengths. Other times, a side length may be missing and must be determined before the perimeter can be calculated. Consider the example below:

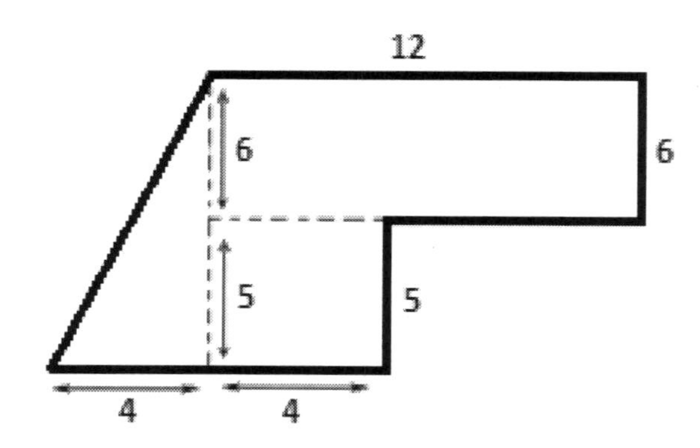

All of the side lengths are provided except for the angled side on the left. Test takers should notice that this is the hypotenuse of a right triangle. The other two sides of the triangle are provided (the base is 4 and the height is 6 + 5 = 11). The Pythagorean Theorem can be used to find the length of the hypotenuse, remembering that $a^2 + b^2 = c^2$.

Substituting the side values provided yields $(4)^2 + (11)^2 = c^2$.

Therefore, $c = \sqrt{16 + 121} = 11.7$

Finally, the perimeter can be found by adding this new side length with the other provided lengths to get the total length around the figure: 4 + 4 + 5 + 6 + 12 + 11.7 = 42.7. Although units are not provided in this figure, remember that reporting units with a measurement is important.

The area of irregular polygons is found by decomposing, or breaking apart, the figure into smaller shapes. When the area of the smaller shapes is determined, the area of the smaller shapes will produce the area of the original figure when added together. Consider the earlier example:

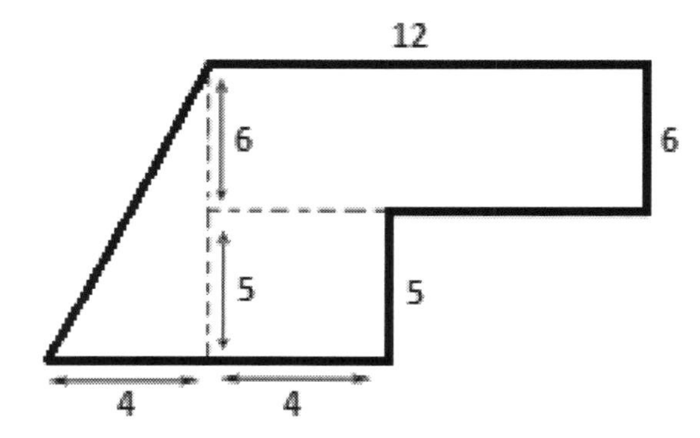

The irregular polygon is decomposed into two rectangles and a triangle. The area of the large rectangles ($A = l \times w \rightarrow A = 12 \times 6$) is 72 square units. The area of the small rectangle is 20 square units ($A = 4 \times 5$). The area of the triangle ($A = \frac{1}{2} \times b \times h \rightarrow A = \frac{1}{2} \times 4 \times 11$) is 22 square units. The sum of the areas of these figures produces the total area of the original polygon: $A = 72 + 20 + 22 \rightarrow A = 114$ square units.

Here's another example:

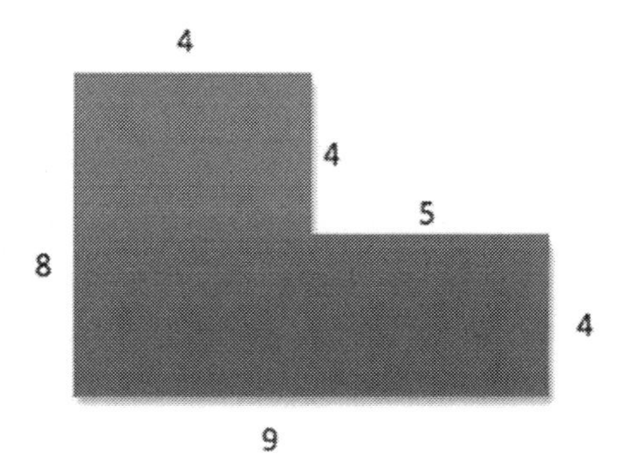

This irregular polygon is decomposed into two rectangles. The area of the large rectangle ($A = l \times w \rightarrow A = 8 \times 4$) is 32 square units. The area of the small rectangle is 20 square units ($A = 4 \times 5$). The sum of the areas of these figures produces the total area of the original polygon: $A = 32 + 20 \rightarrow A = 52$ square units.

Circumference

A circle's perimeter—also known as its *circumference*—is measured by multiplying the *diameter* (the straight line measured from one end to the direct opposite end of the circle) by π, so the formula is $\pi \times d$. This is sometimes expressed by the formula $C = 2 \times \pi \times r$, where r is the radius of the circle. These formulas are equivalent, as the radius equals half of the diameter. The area of a circle is calculated through the formula $A = \pi \times r^2$. The test will indicate either to leave the answer with π attached or to calculate to the nearest decimal place, which means multiplying by 3.14 for π.

Volumes and Surface Areas

Geometry in three dimensions is similar to geometry in two dimensions. The main new feature is that three points now define a unique *plane* that passes through each of them. Three dimensional objects can be made by putting together two dimensional figures in different surfaces. Below, some of the possible three dimensional figures will be provided, along with formulas for their volumes and surface areas.

A rectangular prism is a box whose sides are all rectangles meeting at 90° angles. Such a box has three dimensions: length, width, and height. If the length is *x*, the width is *y*, and the height is *z*, then the volume is given by $V = xyz$.

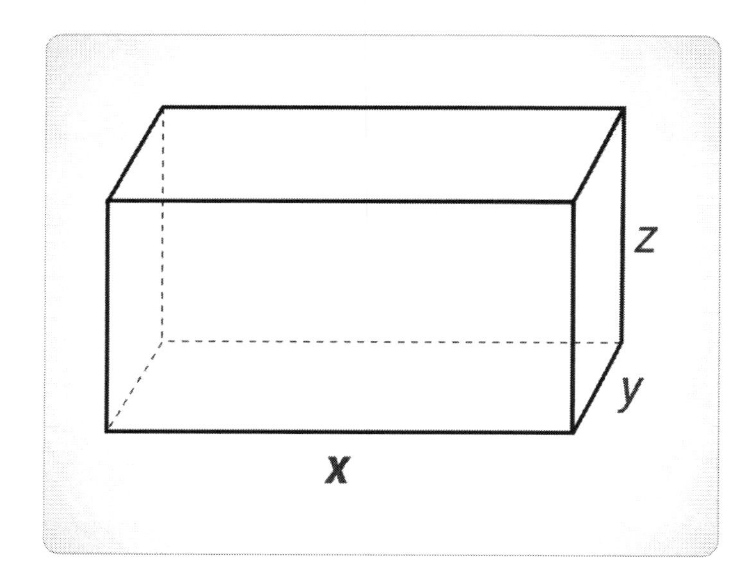

The surface area will be given by computing the surface area of each rectangle and adding them together. There are a total of six rectangles. Two of them have sides of length *x* and *y*, two have sides of length *y* and *z*, and two have sides of length *x* and *z*. Therefore, the total surface area will be given by $SA = 2xy + 2yz + 2xz$.

A *rectangular pyramid* is a figure with a rectangular base and four triangular sides that meet at a single vertex. If the rectangle has sides of length *x* and *y*, then the volume will be given by $V = \frac{1}{3}xyh$.

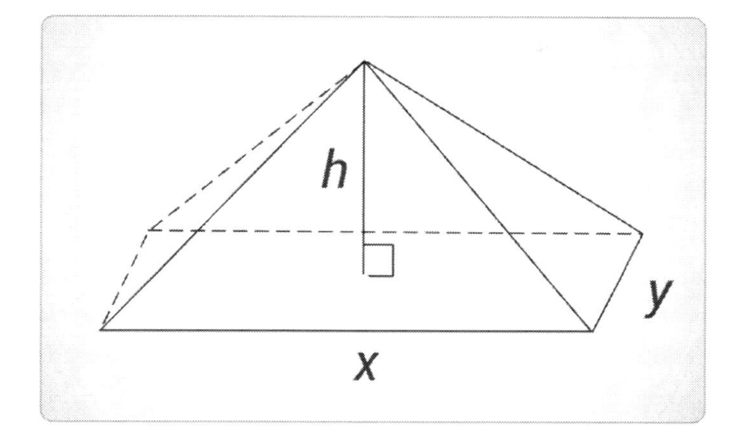

To find the surface area, the dimensions of each triangle must be known. However, these dimensions can differ depending on the problem in question. Therefore, there is no general formula for calculating the total surface area.

A *sphere* is a set of points all of which are equidistant from some central point. It is like a circle, but in three dimensions. The volume of a sphere of radius *r* is given by $V = \frac{4}{3}\pi r^3$. The surface area is given by $A = 4\pi r^2$.

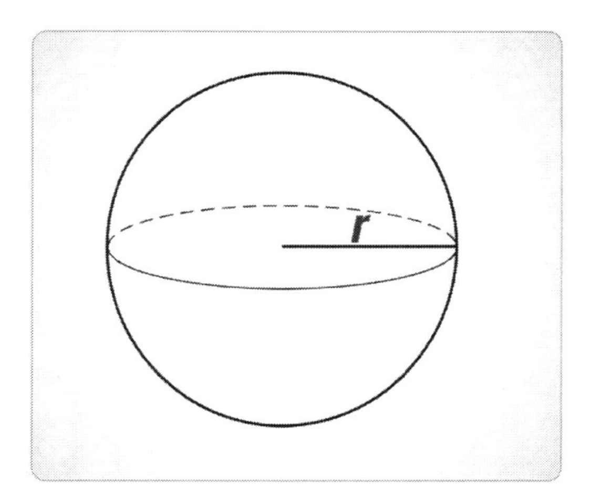

<u>The Pythagorean Theorem</u>

The Pythagorean theorem is an important result in geometry. It states that for right triangles, the sum of the squares of the two shorter sides will be equal to the square of the longest side (also called the *hypotenuse*). The longest side will always be the side opposite to the 90° angle. If this side is called *c*, and the other two sides are *a* and *b*, then the Pythagorean theorem states that $c^2 = a^2 + b^2$. Since lengths are always positive, this also can be written as $c = \sqrt{a^2 + b^2}$.

A diagram to show the parts of a triangle using the Pythagorean theorem is below.

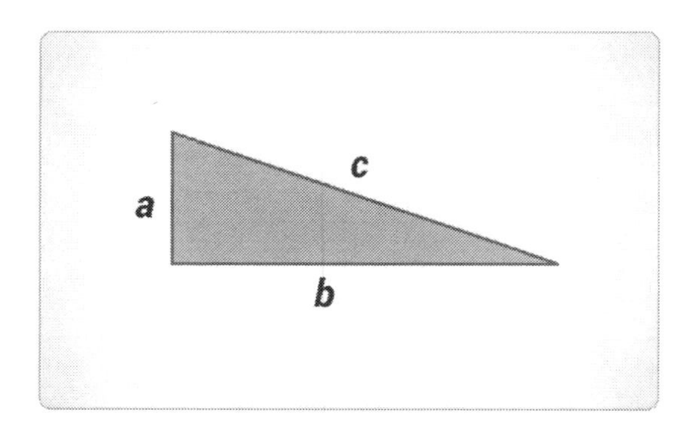

As an example of the theorem, suppose that Shirley has a rectangular field that is 5 feet wide and 12 feet long, and she wants to split it in half using a fence that goes from one corner to the opposite corner. How long will this fence need to be? To figure this out, note that this fence would divide the field into two right triangles, whose hypotenuse will be the fence length. Therefore, the fence length will be given by $\sqrt{5^2 + 12^2} = \sqrt{169} = 13$ feet long.

Data Analysis

Mean, Median, and Mode

Mean

Suppose that you have a set of data points and some description of the general properties of this data need to be found.

The first property that can be defined for this set of data is the *mean*. This is the same as average. To find the mean, add up all the data points, then divide by the total number of data points. For example, suppose that in a class of 10 students, the scores on a test were 50, 60, 65, 65, 75, 80, 85, 85, 90, 100. Therefore, the average test score will be:

$$\frac{50 + 60 + 65 + 65 + 75 + 80 + 85 + 85 + 90 + 100}{10} = 75.5$$

The mean is a useful number if the distribution of data is normal (more on this later), which roughly means that the frequency of different outcomes has a single peak and is roughly equally distributed on both sides of that peak. However, it is less useful in some cases where the data might be split or where there are some *outliers*. Outliers are data points that are far from the rest of the data. For example, suppose there are 11 executives and 90 employees at a company. The executives make $1000 per hour, and the employees make $10 per hour.

Therefore, the average pay rate will be:

$$\frac{\$1000 \cdot 11 + \$10 \cdot 90}{100} = \$119 \textit{ per hour}$$

In this case, this average is not very descriptive since it's not close to the actual pay of the executives *or* the employees.

Median
Another useful measurement is the *median*. In a data set, the median is the point in the middle. The middle refers to the point where half the data comes before it and half comes after, when the data is recorded in numerical order. For instance, these are the speeds of the fastball of a pitcher during the last inning that he pitched (in order from least to greatest):

90, 92, 93, 93, 95, 96, 97, 97, 97

There are nine total numbers, so the middle or *median* number in the 5th one, which is 95.

In cases where the number of data points is an even number, then the average of the two middle points is taken. In the previous example of test scores, the two middle points are 75 and 80. Since there is no single point, the average of these two scores needs to be found. The average is:

$$\frac{75 + 80}{2} = 77.5$$

The median is generally a good value to use if there are a few outliers in the data. It prevents those outliers from affecting the "middle" value as much as when using the mean.

Since an outlier is a data point that is far from most of the other data points in a data set, this means an outlier also is any point that is far from the median of the data set. The outliers can have a substantial effect on the mean of a data set, but usually do not change the median or mode, or do not change them by a large quantity. For example, consider the data set (3, 5, 6, 6, 6, 8). This has a median of 6 and a mode of 6, with a mean of $\frac{34}{6} \approx 5.67$. Now, suppose a new data point of 1000 is added so that the data set is now (3, 5, 6, 6, 6, 8, 1000). This does not change the median or mode, which are both still 6. However, the average is now $\frac{1034}{7}$, which is approximately 147.7. In this case, the median and mode will be better descriptions for most of the data points.

The reason for outliers in a given data set is a complicated problem. It is sometimes the result of an error by the experimenter, but often they are perfectly valid data points that must be taken into consideration.

Mode
One additional measure to define for *X* is the *mode*. This is the data point that appears most frequently. If two or more data points all tie for the most frequent appearance, then each of them is considered a mode. In the case of the test scores, where the numbers were 50, 60, 65, 65, 75, 80, 85, 85, 90, 100, there are two modes: 65 and 85.

<u>Quartiles and Percentiles</u>
The *first quartile* of a set of data *X* refers to the largest value from the first ¼ of the data points. In practice, there are sometimes slightly different definitions that can be used, such as the median of the first half of the data points (excluding the median itself if there are an odd number of data points). The term also has a slightly different use: when it is said that a data point lies *in the first quartile*, it means it is less than or equal to the median of the first half of the data points. Conversely, if it lies *at* the first quartile, then it is equal to the first quartile.

When it is said that a data point lies in the *second quartile*, it means it is between the first quartile and the median.

The *third quartile* refers to data that lies between ½ and ¾ of the way through the data set. Again, there are various methods for defining this precisely, but the simplest way is to include all of the data that lie between the overall median and the median of the top half of the data.

Data that lies in the *fourth quartile* refers to all of the data above the third quartile.

Percentiles may be defined in a similar manner to quartiles. Generally, this is defined in the following manner:

If a data point lies *in the n-th percentile*, it means that it lies in the range of the first *n*% of the data.

If a data point lies *at* the *n*-th percentile, then it means that *n*% of the data lies below this data point.

Standard Deviation

Given a data set *X* consisting of data points $(x_1, x_2, x_3, \ldots x_n)$, the *variance* of *X* is defined to be $\frac{\sum_{i=1}^{n}(x_i - \bar{X})^2}{n}$. This means that the variance of *X* is the average of the squares of the differences between each data point and the mean of *X*.

Given a data set *X* consisting of data points $(x_1, x_2, x_3, \ldots x_n)$, the *standard deviation* of *X* is defined to be $s_x = \sqrt{\frac{\sum_{i=1}^{n}(x_i - \bar{X})^2}{n}}$. In other words, the standard deviation is the square root of the variance.

Both the variance and the standard deviation are measures of how much the data tend to be spread out. When the standard deviation is low, the data points are mostly clustered around the mean. When the standard deviation is high, it generally indicates that the data are quite spread out, or else that there are a few substantial outliers.

As a simple example, compute the standard deviation for the data set (1, 3, 3, 5). The first step is to compute the mean, which is $\frac{1+3+3+5}{4} = \frac{12}{4} = 3$. Next, the variance of *X* is found with the formula: $\sum_{i=1}^{4}(x_i - \bar{X})^2 = (1-3)^2 + (3-3)^2 + (5-3)^2 = -2^2 + 0^2 + 0^2 + 2^2 = 8$. Therefore, the variance is $\frac{8}{4} = 2$. Taking the square root, the standard deviation is found to be $\sqrt{2}$.

Note that the standard deviation only depends upon the mean, not upon the median or mode(s). Generally, if there are multiple modes that are far apart from one another, the standard deviation will be high. A high standard deviation does not always mean there are multiple modes, however.

Interpretation of Graphs

Data can be represented in many ways including picture graphs, bar graphs, line plots, and tally charts. It is important to be able to organize the data into categories that could be represented using one of these methods. Equally important is the ability to read these types of diagrams and interpret their meaning.

A *picture graph* is a diagram that shows pictorial representation of data being discussed. The symbols used can represent a certain number of objects. Notice how each fruit symbol in the following graph represents a count of two fruits. One drawback of picture graphs is that they can be less accurate if each symbol represents a large number. For example, if each banana symbol represented ten bananas, and students consumed 22 bananas, it may be challenging to draw and interpret two and one-fifth bananas as a frequency count of 22.

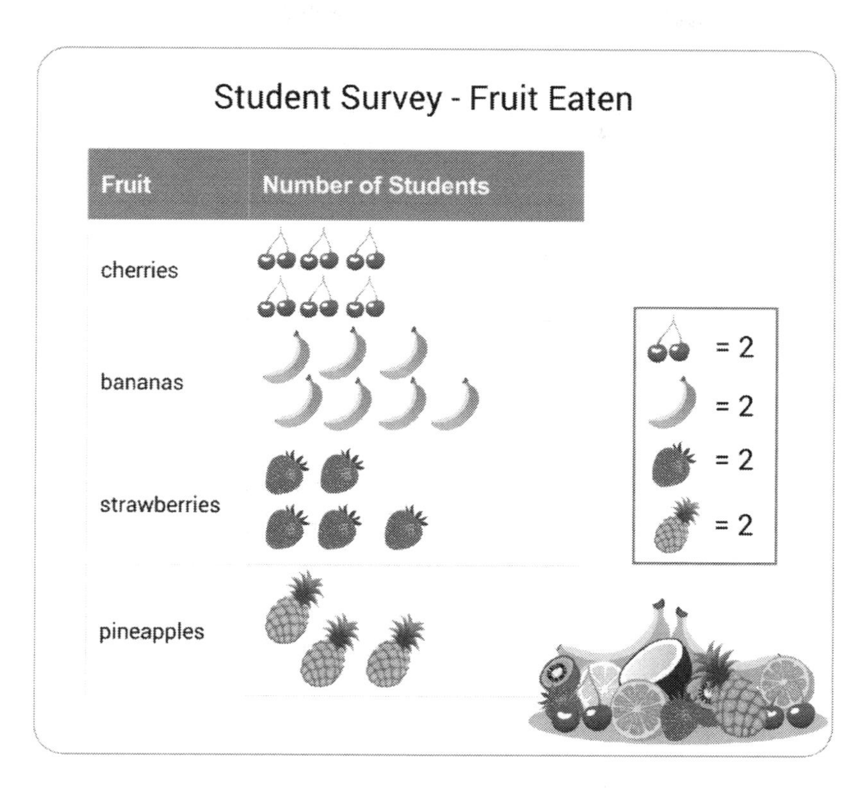

A *bar graph* is a diagram in which the quantity of items within a specific classification is represented by the height of a rectangle. Each type of classification is represented by a rectangle of equal width. Here is an example of a bar graph:

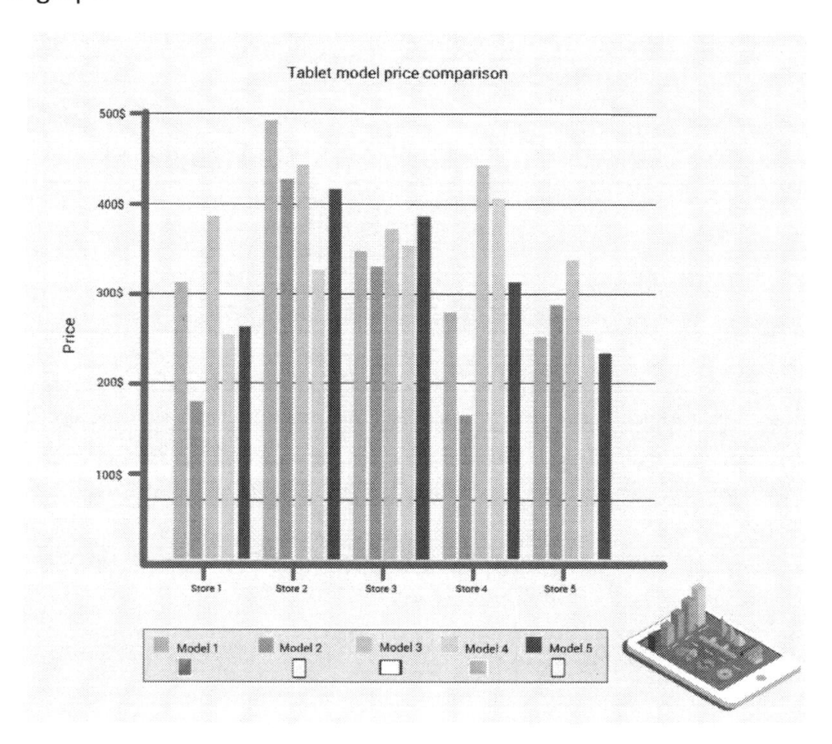

A *line plot* is a diagram that shows quantity of data along a number line. It is a quick way to record data in a structure similar to a bar graph without needing to do the required shading of a bar graph. Here is an example of a line plot:

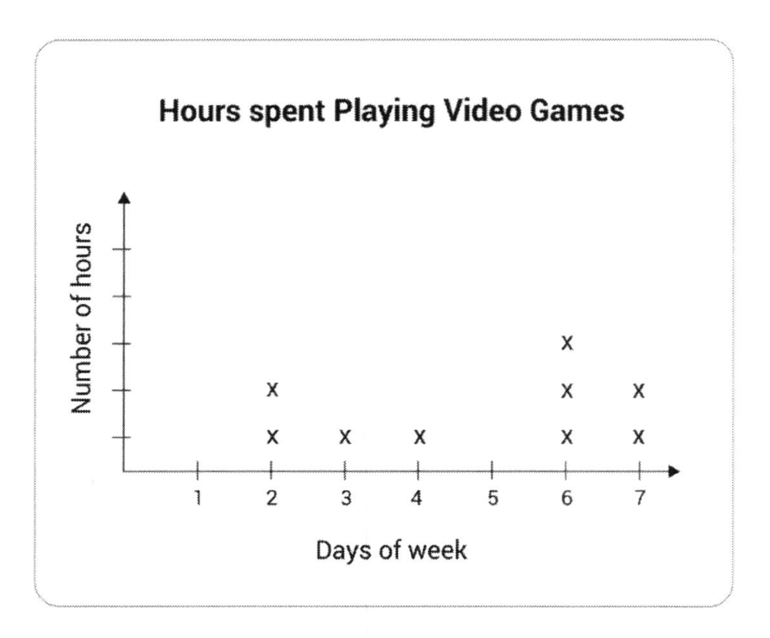

A *tally chart* is a diagram in which tally marks are utilized to represent data. Tally marks are a means of showing a quantity of objects within a specific classification. Here is an example of a tally chart:

Number of days with rain	Number of weeks
0	||
1	𝍸
2	𝍸
3	𝍸
4	𝍸 𝍸 𝍸 ||||
5	𝍸 |
6	𝍸 |
7	||||

Data is often recorded using fractions, such as half a mile, and understanding fractions is critical because of their popular use in real-world applications. Also, it is extremely important to label values with their units when using data. For example, regarding length, the number 2 is meaningless unless it is attached to a unit. Writing 2 cm shows that the number refers to the length of an object.

A circle graph, also called a pie chart, shows categorical data with each category representing a percentage of the whole data set. To make a circle graph, the percent of the data set for each category must be determined. To do so, the frequency of the category is divided by the total number of data points and converted to a percent. For example, if 80 people were asked what their favorite sport is and

20 responded basketball, basketball makes up 25% of the data ($\frac{20}{80}$ =.25=25%). Each category in a data set is represented by a *slice* of the circle proportionate to its percentage of the whole.

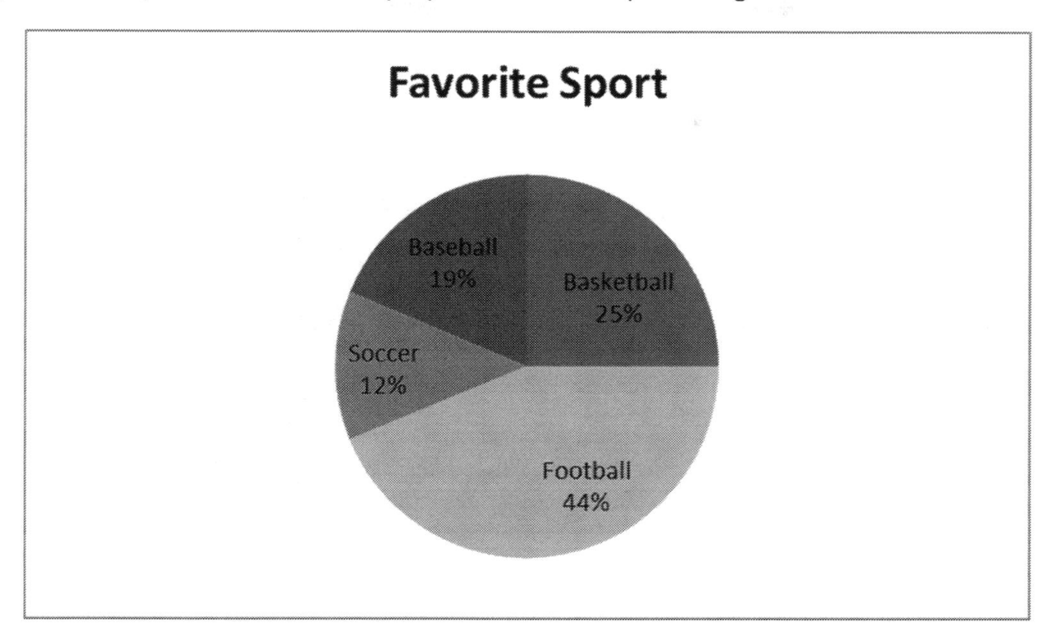

A scatter plot displays the relationship between two variables. Values for the independent variable, typically denoted by *x*, are paired with values for the dependent variable, typically denoted by *y*. Each set of corresponding values are written as an ordered pair (*x*, *y*). To construct the graph, a coordinate grid is labeled with the *x*-axis representing the independent variable and the *y*-axis representing the dependent variable. Each ordered pair is graphed.

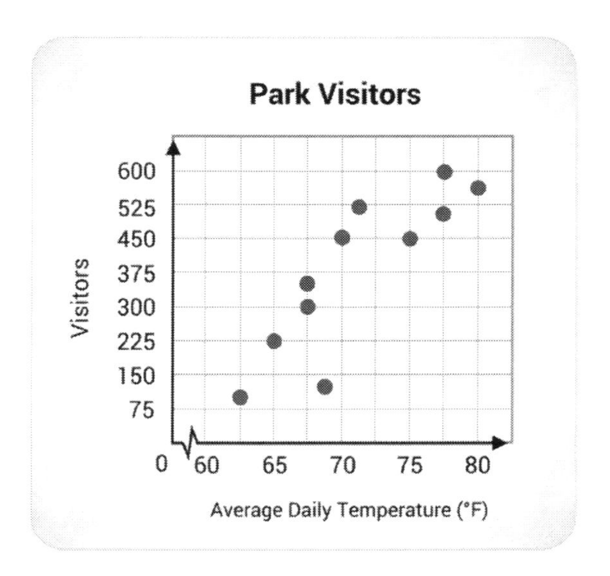

Like a scatter plot, a line graph compares two variables that change continuously, typically over time. Paired data values (ordered pair) are plotted on a coordinate grid with the *x*- and *y*-axis representing the two variables. A line is drawn from each point to the next, going from left to right. A double line graph simply displays two sets of data that contain values for the same two variables. The double line graph below displays the profit for given years (two variables) for Company A and Company B (two data sets).

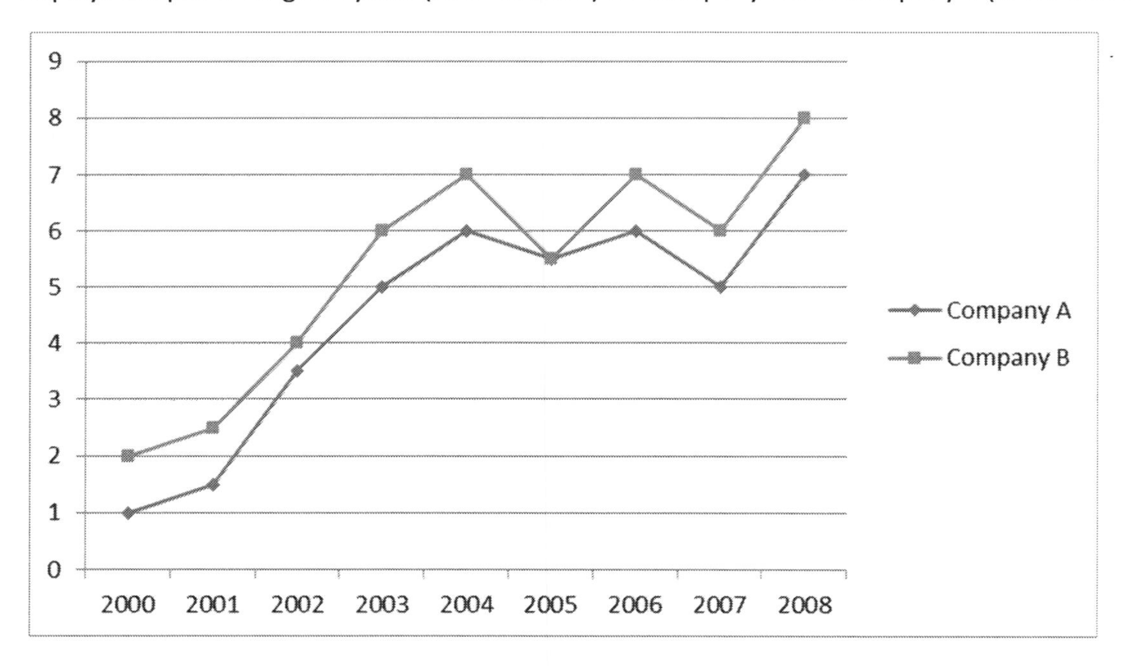

Choosing the appropriate graph to display a data set depends on what type of data is included in the set and what information must be shown. Histograms and box plots can be used for data sets consisting of individual values across a wide range. Examples include test scores and incomes. Histograms and box plots will indicate the center, spread, range, and outliers of a data set. A histogram will show the shape of the data set, while a box plot will divide the set into quartiles (25% increments), allowing for comparison between a given value and the entire set.

Scatter plots and line graphs can be used to display data consisting of two variables. Examples include height and weight, or distance and time. A correlation between the variables is determined by examining the points on the graph. Line graphs are used if each value for one variable pairs with a distinct value for the other variable. Line graphs show relationships between variables.

Probabilities

Given a set of possible outcomes *X*, a *probability distribution* on *X* is a function that assigns a probability to each possible outcome. If the outcomes are $(x_1, x_2, x_3, \dots x_n)$, and the probability distribution is *p*, then the following rules are applied.

- $0 \le p(x_i) \le 1$, for any i.
- $\sum_{i=1}^{n} p(x_i) = 1$.

In other words, the probability of a given outcome must be between zero and 1, while the total probability must be 1.

If $p(x_i)$ is constant, then this is called a *uniform probability distribution*, and $p(x_i) = \frac{1}{n}$. For example, on a six-sided die, the probability of each of the six outcomes will be $\frac{1}{6}$.

If seeking the probability of an outcome occurring in some specific range *A* of possible outcomes, written $P(A)$, the probabilities for each outcome in that range are added together. For example, consider a six-sided die, and figure the probability of getting a 3 or lower when it is rolled. The possible rolls are 1, 2, 3, 4, 5, and 6. So, to get a 3 or lower, a roll of 1, 2, or 3 must be completed. The probabilities of each of these is $\frac{1}{6}$, so add these to get $p(1) + p(2) + p(3) = \frac{1}{6} + \frac{1}{6} + \frac{1}{6} = \frac{1}{2}$.

<u>Probabilities of Simple and Compound Events and of Independent and Dependent Events</u>

Simple and Compound Events

A *simple event* consists of only one outcome. The most popular simple event is flipping a coin, which results in either heads or tails. A compound event results in more than one outcome and consists of more than one simple event. An example of a compound event is flipping a coin while tossing a die. The result is either heads or tails on the coin and a number from one to six on the die. The probability of a simple event is calculated by dividing the number of possible outcomes by the total number of outcomes. Therefore, the probability of obtaining heads on a coin is $^1/_2$, and the probability of rolling a 6 on a die is $^1/_6$. The probability of compound events is calculated using the basic idea of the probability of simple events. If the two events are independent, the probability of one outcome is equal to the product of the probabilities of each simple event. For example, the probability of obtaining heads on a coin and rolling a 6 is equal to $^1/_2 \times {}^1/_6 = {}^1/_{12}$. The probability of either A or B occurring is equal to the sum of the probabilities minus the probability that both A and B will occur. Therefore, the probability of obtaining either heads on a coin or rolling a 6 on a die is $^1/_2 + {}^1/_6 - {}^1/_{12} = {}^7/_{12}$. The two events aren't mutually exclusive because they can happen at the same time. If two events are mutually exclusive, and the probability of both events occurring at the same time is zero, the probability of event A or B occurring equals the sum of both probabilities. An example of calculating the probability of two mutually exclusive events is determining the probability of pulling a king or a queen from a deck of cards. The two events cannot occur at the same time.

Measuring Probabilities with Two-Way Frequency Tables

When measuring event probabilities, two-way frequency tables can be used to report the raw data and then used to calculate probabilities. If the frequency tables are translated into relative frequency tables, the probabilities presented in the table can be plugged directly into the formulas for conditional probabilities. By plugging in the correct frequencies, the data from the table can be used to determine if events are independent or dependent.

Differing Probabilities

The probability that event A occurs differs from the probability that event A occurs given B. When working within a given model, it's important to note the difference. $P(A|B)$ is determined using the formula $P(A|B) = \frac{P\,(A \text{ and } B)}{P(B)}$ and represents the total number of A's outcomes left that could occur after B occurs. $P(A)$ can be calculated without any regard for B. For example, the probability of a student finding a parking spot on a busy campus is different once class is in session.

The Addition Rule

The probability of event A or B occurring isn't equal to the sum of each individual probability. The probability that both events can occur at the same time must be subtracted from this total. This idea is shown in the *addition rule*: $P(A \text{ or } B) = P(A) + P(B) - P(A \text{ and } B)$. The addition rule is another way to determine the probability of compound events that aren't mutually exclusive. If the events are mutually exclusive, the probability of both A and B occurring at the same time is 0.

Uniform and Non-Uniform Probability Models

A *uniform probability model* is one where each outcome has an equal chance of occurring, such as the probabilities of rolling each side of a die. A *non-uniform probability model* is one where each outcome has an unequal chance of occurring. In a uniform probability model, the conditional probability formulas for $P(B|A)$ and $P(A|B)$ can be multiplied by their respective denominators to obtain two formulas for $P(A \text{ and } B)$. Therefore, the multiplication rule is derived as $P(A \text{ and } B) = P(A)P(B|A) = P(B)P(A|B)$. In a model, if the probability of either individual event is known and the corresponding conditional probability is known, the multiplication rule allows the probability of the joint occurrence of A and B to be calculated.

Binomial Experiments

In statistics, a *binomial experiment* is an experiment that has the following properties. The experiment consists of *n* repeated trial that can each have only one of two outcomes. It can be either a success or a failure. The probability of success, *p*, is the same in every trial. Each trial is also independent of all other trials. An example of a binomial experiment is rolling a die 10 times with the goal of rolling a 5. Rolling a 5 is a success while any other value is a failure. In this experiment, the probability of rolling a 5 is $\frac{1}{6}$. In any binomial experiment, x is the number of resulting successes, n is the number of trials, p is the probability of success in each trial, and $q = 1 - p$ is the probability of failure within each trial. The probability of obtaining x successes within n trials is:

$$P(X = x) = \frac{n!}{x!\,(n-x)!}p^x(1-p)^{n-x}$$

With the following being the *binomial coefficient*:

$$\binom{n}{x} = \frac{n!}{x!\,(n-x)!}$$

Within this calculation, $n!$ is n factorial that's defined as:

$$n \cdot (n-1) \cdot (n-2) \dots 1$$

Let's look at the probability of obtaining 2 rolls of a 5 out of the 10 rolls.

Start with $P(X = 2)$, where 2 is the number of successes. Then fill in the rest of the formula with what is known, *n*=10, *x*=2, *p*=1/6, *q*=5/6:

$$P(X = 2) = \left(\frac{10!}{2!\,(10-2)!}\right)\left(\frac{1}{6}\right)^2\left(1-\frac{1}{6}\right)^{10-2}$$

Which simplifies to:

$$P(X = 2) = \left(\frac{10!}{2!\,8!}\right)\left(\frac{1}{6}\right)^2\left(\frac{5}{6}\right)^8$$

Then solve to get:

$$P(X = 2) = \left(\frac{3628800}{80640}\right)(.0277)(.2325) = .2898$$

Conditional Probability

Sample Subsets

A sample can be broken up into subsets that are smaller parts of the whole. For example, consider a sample population of females. The sample can be divided into smaller subsets based on the characteristics of each female. There can be a group of females with brown hair and a group of females that wear glasses. There also can be a group of females that have brown hair *and* wear glasses. This "and" relates to the *intersection* of the two separate groups of brunettes and those with glasses. Every female in that intersection group has both characteristics. Similarly, there also can be a group of females that either have brown hair *or* wear glasses. The "or" relates to the union of the two separate groups of brunettes and glasses. Every female in this group has at least one of the characteristics. Finally, the group of females who do *not* wearing glasses can be discussed. This "not" relates to the *complement* of the glass-wearing group. No one in the complement has glasses. *Venn diagrams* are useful in highlighting these ideas. When discussing statistical experiments, this idea can also relate to events instead of characteristics.

Verifying Independent Events

Two events aren't always independent. For examples, females with glasses and brown hair aren't independent characteristics. There definitely can be overlap because females with brown hair can wear glasses. Also, two events that exist at the same time don't have to have a relationship. For example, even if all females in a given sample are wearing glasses, the characteristics aren't related. In this case, the probability of a brunette wearing glasses is equal to the probability of a female being a brunette multiplied by the probability of a female wearing glasses. This mathematical test of $P(A \cap B) = P(A)P(B)$ verifies that two events are independent.

Conditional Probability

Conditional probability is the probability that event A will happen given that event B has already occurred. An example of this is calculating the probability that a person will eat dessert once they have eaten dinner. This is different than calculating the probability of a person just eating dessert. The formula for the conditional probability of event A occurring given B is $P(A|B) = \frac{P\ (A \text{ and } B)}{P(B)}$, and it's defined to be the probability of both A and B occurring divided by the probability of event B occurring. If A and B are independent, then the probability of both A and B occurring is equal to $P(A)P(B)$, so $P(A|B)$ reduces to just $P(A)$. This means that A and B have no relationship, and the probability of A occurring is the same as the conditional probability of A occurring given B. Similarly, $P(B|A) = \frac{P\ (B \text{ and } A\ )}{P(A)} = P(B)$ if A and B are independent.

Independent Versus Related Events

To summarize, conditional probability is the probability that an event occurs given that another event has happened. If the two events are related, the probability that the second event will occur changes if the other event has happened. However, if the two events aren't related and are therefore independent, the occurrence of the first event won't impact the probability of the second event occurring.

Combinations and Permutations

There are many counting techniques that can help solve problems involving counting possibilities. For example, the *Addition Principle* states that if there are m choices from Group 1 and n choices from Group 2, then $n + m$ is the total number of choices possible from Groups 1 and 2. For this to be true, the groups can't have any choices in common. The *Multiplication Principle* states that if Process 1 can be completed n ways and Process 2 can be completed m ways, the total number of ways to complete both

Process 1 and Process 2 is $n \times m$. For this rule to be used, both processes must be independent of each other. Counting techniques also involve permutations. A *permutation* is an arrangement of elements in a set for which order must be considered. For example, if three letters from the alphabet are chosen, ABC and BAC are two different permutations. The multiplication rule can be used to determine the total number of possibilities. If each letter can't be selected twice, the total number of possibilities is $26 \times 25 \times 24 = 15{,}600$. A formula can also be used to calculate this total. In general, the notation $P(n, r)$ represents the number of ways to arrange r objects from a set of n and, the formula is $P(n, r) = \frac{n!}{(n-r)!}$. In the previous example, $P(26, 3) = \frac{26!}{23!} = 15{,}600$.

Contrasting permutations, a *combination* is an arrangement of elements in which order doesn't matter. In this case, ABC and BAC are the same combination. In the previous scenario, there are six permutations that represent each single combination. Therefore, the total number of possible combinations is $15{,}600 \div 6 = 2{,}600$. In general, $C(n, r)$ represents the total number of combinations of n items selected r at a time where order doesn't matter, and the formula is $C(n, r) = \frac{n!}{(n-r)!\,r!}$. Therefore, the following relationship exists between permutations and combinations: $C(n, r) = \frac{P(n,r)}{n!} = \frac{P(n,r)}{P(r,r)}$.

GRE Quantitative Reasoning Practice Test #1

For each of questions 1-8, compare Quantity A to Quantity B, using additional information presented above the two quantities. Select one of the following answer choices for each question:

a. Quantity A is greater
b. Quantity B is greater
c. The two quantities are equal
d. The relationship cannot be determined from the information given.

1. h is an integer in the following mathematical series: 4, h, 19, 39, 79

Quantity A	Quantity B
The value of h	9

a.
b.
c.
d.

2.

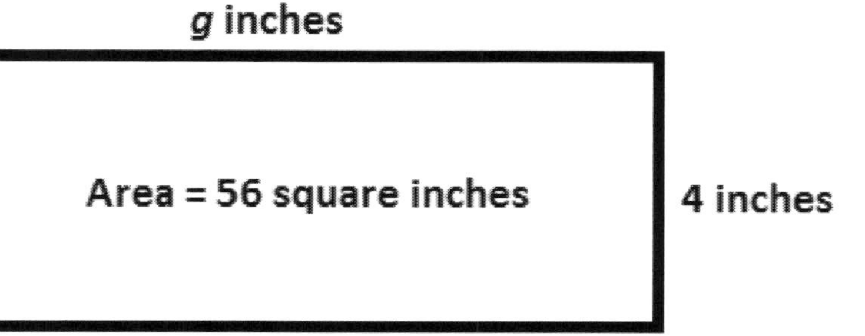

Quantity A	Quantity B
The value of g	13

a.
b.
c.
d.

3. $4x - 12 = -2x$

Quantity A	Quantity B
The value of x	3

a.
b.
c.
d.

4.

Jimmy	Steve
7 red marbles	6 green marbles
8 blue marbles	4 blue marbles

Quantity A
All of Jimmy's marbles divided by all of Steve's marbles

Quantity B
Jimmy's blue marbles divided by Steve's green marbles

a.
b.
c.
d.

5.

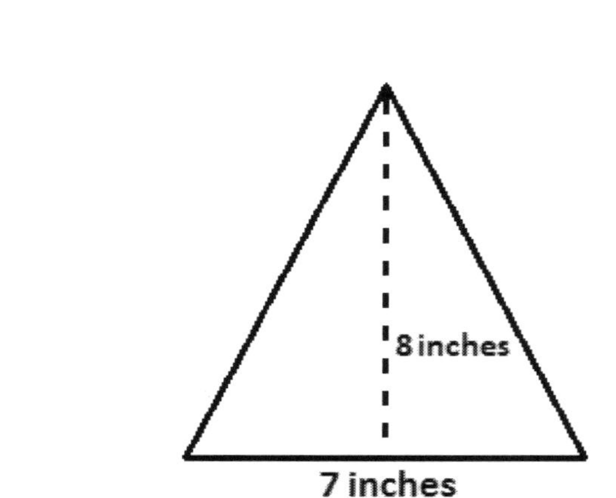

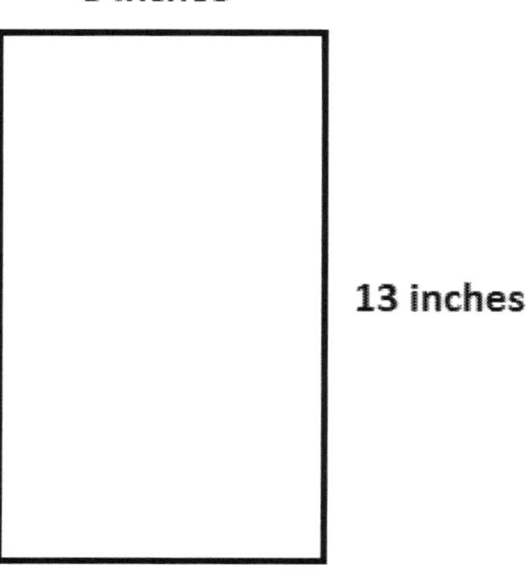

Quantity A
7 times the area of the triangle

Quantity B
2 times the area of the rectangle

a.
b.
c.
d.

6. Truck A drives 1236 yards and truck B drives 3680 feet.

Quantity A
The distance that truck *A* drove

Quantity B
The distance that truck *B* drove

a.
b.
c.
d.

7. $x > 6 > z$

Quantity A	Quantity B
$x + z$	$x - 6$

a.
b.
c.
d.

8. There are 16 rocks in a bag. 12 of them are smooth and 4 of them are rough.

Quantity A	Quantity B
The probability of choosing a rough rock	$\frac{2}{8}$

a.
b.
c.
d.

Questions 9-20 have several different formats. Unless otherwise directed, select a single answer choice for each question. For Numeric Entry questions, follow the instructions below.

Numeric Entry Questions

Enter your answer in the box(es) below the question.

- Your answer may be an integer, a decimal, a fraction, and it may be negative.
- If a question asks for a fraction, there will be two boxes. One is for the numerator and one is for the denominator.
- Equivalent forms of the value, such as 1.5 and 1.50, are all correct. Fractions do not need to be reduced to lowest terms.
- Enter the exact answer unless your question asks you to round your answer.

9. Alan currently weighs 200 pounds, but he wants to lose weight to get down to 175 pounds. What is this difference in kilograms? (1 pound is approximately equal to 0.45 kilograms.)
a. 9 kg
b. 11.25 kg
c. 78.75 kg
d. 90 kg
e. 25 kg

10. Johnny earns $2334.50 from his job each month. He pays $1437 for monthly expenses. Johnny is planning a vacation in 3 months' time that he estimates will cost $1750 total. How much will Johnny have left over from three months' of saving once he pays for his vacation?
a. $948.50
b. $584.50
c. $852.50
d. $942.50
e. $848.50

11. What is $\frac{420}{98}$ rounded to the nearest integer?

a. 3
b. 4
c. 5
d. 6
e. 7

For the following question, select all that apply:

12. Five students took a test. Jenny scored the highest with a 94. James scored the lowest with a 79. Hector scored lower than Jenny, but higher than Sam. Sam scored lower than Mary who scored an 84. Which of the following statements must be true?

a. There were 3 people who scored higher than Sam.
b. The median test score was an 84.
c. The average test score must be higher than 82.
d. Jenny is the only student who scored above 90.
e. Hector scored lower than Mary.

13. The following graph compares the various test scores of the top three students in each of these teacher's classes. Based on the graph, which teacher's students had the smallest range of test scores?

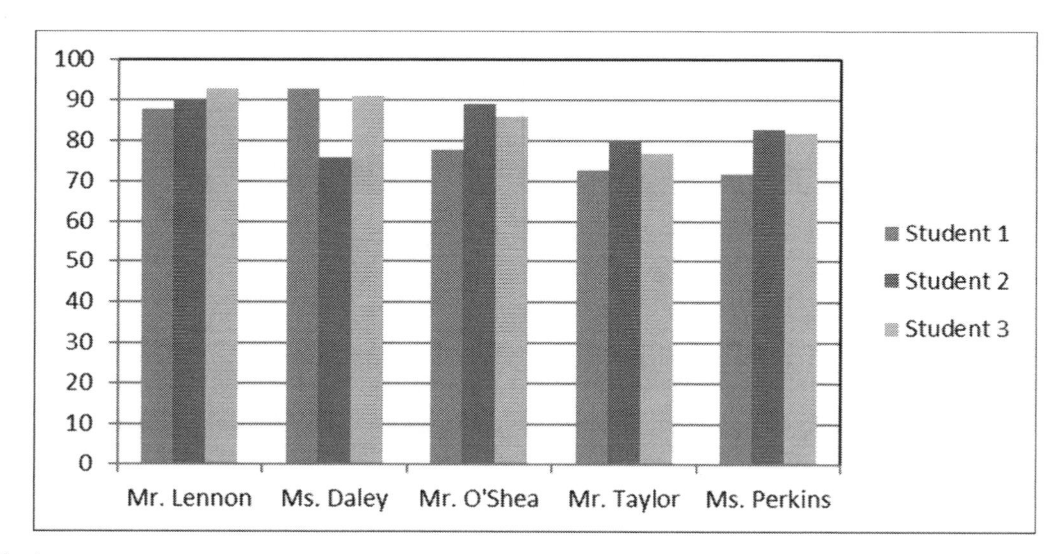

a. Mr. Lennon
b. Mr. O'Shea
c. Mr. Taylor
d. Ms. Daley
e. Ms. Perkins

For the following question, enter your answer in the box:

14. A local candy store reports that of the 100 customers that bought suckers, 35 of them bought cherry. What is the probability of selecting 2 customers simultaneously at random that both purchased a cherry sucker?

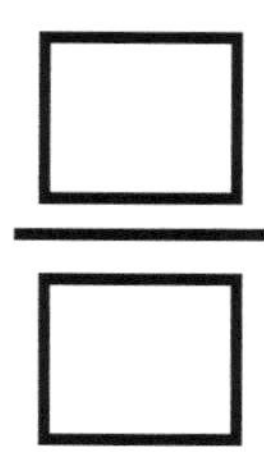

15. Simplify the following expression:

$$4\frac{2}{3} - 3\frac{4}{9}$$

a. $1\frac{1}{3}$
b. $1\frac{2}{9}$
c. 1
d. $1\frac{2}{3}$
e. $1\frac{3}{4}$

16. The width of a rectangular house is 22 feet. What is the perimeter of this house if it has the same area as a house that is 33 feet wide and 50 feet long?
a. 184 feet
b. 200 feet
c. 192 feet
d. 206 feet
e. 194 feet

17. Using the following diagram, what is the total circumference, rounding to the nearest decimal place?

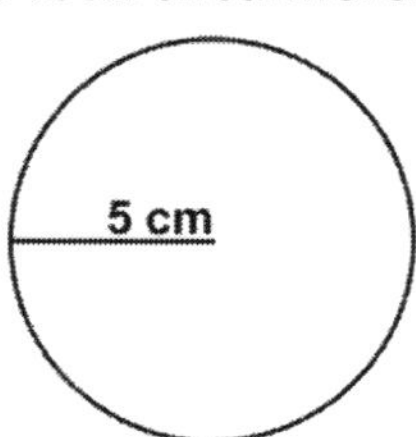

a. 25.0 cm
b. 15.7 cm
c. 78.5 cm
d. 31.4 cm
e. 75.6 cm

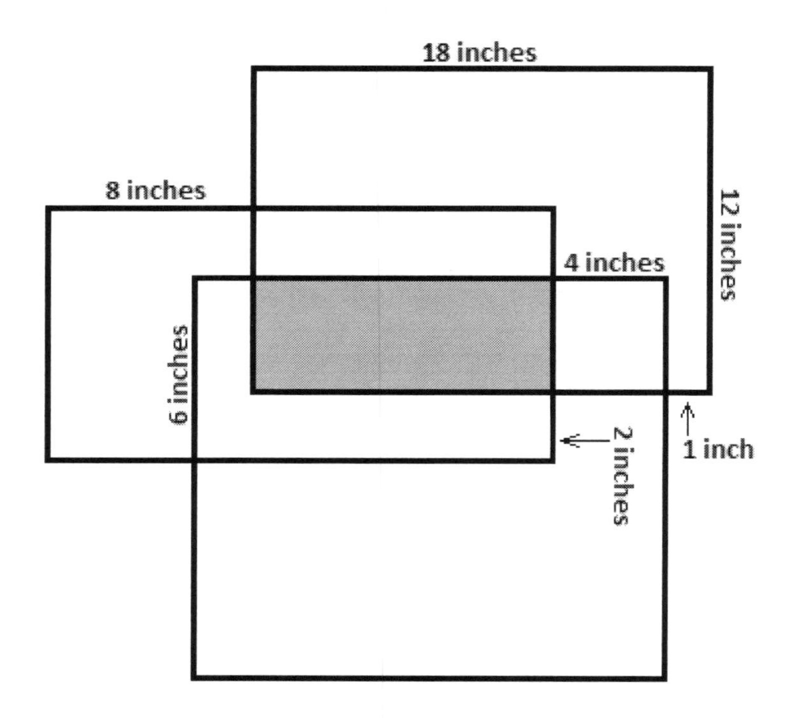

18. In the figure above, what is the area of the shaded region?
    a. 48 sq. inches
    b. 52 sq. inches
    c. 44 sq. inches
    d. 56 sq. inches
    e. 46 sq. inches

19. If $3x = 6y = -2z = 24$, then what does $4xy + z$ equal?
    a. 116
    b. 130
    c. 84
    d. 108
    e. 98

20. If $n = 2^2$, and $m = n^2$, then m^n equals?
    a. 2^{12}
    b. 2^{10}
    c. 2^{18}
    d. 2^{16}
    e. 2^{20}

Answer Explanations Test #1

1. C: The equation that produces this series is $2x + 1$. This gives $2(4) + 1 = 9, 2(9) + 1 = 19$, and so on. This means that the value of *h* in the series is 9, so *Quantity A* and *Quantity B* are equal.

2. A: The value of *g* can be found using the formula for area of a rectangle ($A = l * w$). So, $56 = g * 4$, and $g = 14$. This means that *Quantity A* is greater than *Quantity B*.

3. B: The first step is to solve for x. For this equation that is $4x - 12 = -2x, 6x - 12 = 0, 6x = 12, x = 2$. Since the value of x is 2 and *Quantity B* is 3, it means that *Quantity B* is greater.

4. A: The first step here is to solve each of the ratios. The first ratio is all of Jimmy's marbles divided by all of Steve's marbles. This gives $\frac{15}{10} = \frac{3}{2}$. The second ratio is all of Jimmy's blue marbles divided by all of Steve's green marbles. This gives $\frac{8}{6} = \frac{4}{3}$. Since $\frac{3}{2}$ is greater than $\frac{4}{3}$, *Quantity A* is greater.

5. B: First, find the area of both figures. The area of the triangle is $\frac{1}{2}(7) * 8 = 28$ square inches. The area of the rectangle is $13 * 8 = 104$ square inches. So, 7 times the area of the triangle would be 196 square inches, and 2 times the area of the rectangle would be 208 square inches. This means that *Quantity B* is greater.

6. A: First, convert the distance that *Truck A* drove to feet. This is $1{,}236 * 3 = 3{,}708$ feet. This mean that *Truck A* drove further than *Truck B*. So, *Quantity A* is greater than *Quantity B*.

7. D: There is not enough information to determine which quantity is greater. Either quantity could be greater for given values of x and z. For example, if $x = 7$ and $z = 1$, then *Quantity A* is 8 and *Quantity B* is 1. If, $x = 10$ and $z = -8$, then *Quantity A* is 2 and *Quantity B* is 4.

8. C: The probability of choosing a rough rock is $\frac{4}{16}$. This is equal to $\frac{2}{8}$.

9. B: Using the conversion rate, multiply the projected weight loss of 25 lbs. by 0.45 $\frac{kg}{lb}$ to get the amount in kilograms (11.25 kg).

10. D: First, subtract $1437 from $2334.50 to find Johnny's monthly savings; this equals $897.50. Then, multiply this amount by 3 to find out how much he will have (in three months) before he pays for his vacation: this equals $2692.50. Finally, subtract the cost of the vacation ($1750) from this total to find how much Johnny will have left: $942.50.

11. B: Dividing by 98 can be approximated by dividing by 100, which would mean shifting the decimal point of the numerator to the left by 2. The result is 4.2 which rounds to 4.

12. A & C: It can be determined from reading the information given that Jenny, Hector, and Mary scored higher than Sam, so Choice *A* is correct. There is no relation provided between Hector and Mary's scores. Given that Mary could have scored higher or lower than Hector, it cannot be determined if her score is the median, so Choice *B* is incorrect. Three of the test scores are given and since James has to be the lowest at 79, the lowest that the two missing scores could be are 80 and 81. This means that the lowest that the average could be is 83.6. So, Choice *C* is correct. With the information given, it is possible that Hector scored above 90, so Choice *D* is incorrect. There is no relation given between Hector and

Mary's scores. This means that Hector could have scored higher or lower than Mary. So, Choice *E* is incorrect.

13. A: To calculate the range in a set of data, subtract the lowest value from the highest value. In this graph, the range of Mr. Lennon's students is 5, which can be seen physically in the graph as having the smallest difference between the highest value and the lowest value compared with the other teachers.

14. $\frac{119}{990}$: The probability of choosing two customers simultaneously is the same as choosing one and then choosing a second without putting the first back into the pool of customers. This means that the probability of choosing a customer who bought cherry is $\frac{35}{100}$. Then without placing them back in the pool, it would be $\frac{34}{99}$. So, the probability of choosing 2 customers simultaneously that both bought cherry would be $\frac{35}{100} \times \frac{34}{99} = \frac{1,190}{9,900} = \frac{119}{990}$.

15. B: Simplify each mixed number of the problem into a fraction by multiplying the denominator by the whole number and adding the numerator:

$$\frac{14}{3} - \frac{31}{9}$$

Since the first denominator is a multiple of the second, simplify it further by multiplying both the numerator and denominator of the first expression by 3 so that the denominators of the fractions are equal.

$$\frac{42}{9} - \frac{31}{9} = \frac{11}{9}$$

Simplifying this further, divide the numerator 11 by the denominator 9; this leaves 1 with a remainder of 2. To write this as a mixed number, place the remainder over the denominator, resulting in $1\frac{2}{9}$, Choice *B*.

16. E: First, find the area of the second house. The area is $A = l \; x \; w = 33 \; \times \; 50 = 1{,}650$ square feet. Then use the area formula to determine what length gives the first house an area of 1,650 square feet. So, $1{,}650 = 22 \; \times \; l, l = \frac{1{,}650}{22} = 75$ feet. Then, use the formula for perimeter to get $75 + 75 + 22 + 22 = 194$ feet.

17. D: To calculate the circumference of a circle, use the formula $2\pi r$, where r equals the radius, or half of the diameter, of the circle and $\pi = 3.14 \ldots$. Substitute the given information, $2\pi 5 = 31.4 \ldots$, which is answer *D*.

18. B: This can be determined by finding the length and width of the shaded region. The length can be found using the length of the top rectangle which is 18 inches, then subtracting the extra length of 4 inches and 1 inch. This means the length of the shaded region is 13 inches. Next, the width can be determined using the 6 inch measurement and subtracting the 2 inch measurement. This means that the width is 4 inches. Thus, the area is $13 \; \times \; 4 = 52$ sq. inches.

19. A: First solve for *x, y,* and *z.* So, $3x = 24, x = 8, 6y = 24, y = 4$, and $-2z = 24, z = -12$. This means the equation would be $4(8)(4) + (-12)$, which equals 116.

20. D: If $n = 2^2, n = 4$, and $m = 4^2 = 16$. This means that $m^n = 16^4$. This is the same as 2^{16}.

GRE Quantitative Reasoning Practice Test #2

For each of questions 1-8, compare Quantity A to Quantity B, using additional information presented above the two quantities. Select one of the following answer choices for each question:

a. Quantity A is greater
b. Quantity B is greater
c. The two quantities are equal
d. The relationship cannot be determined from the information given.

1. 85% of 20 = a

Quantity A	Quantity B
The value of a	16

a.
b.
c.
d.

2. The average of six numbers is 6, and the sum of five of the six numbers is 25.

Quantity A	Quantity B
The value of the sixth number	12

a.
b.
c.
d.

3. $\frac{5}{2} \div \frac{1}{3} = n$

Quantity A	Quantity B
The value of n	7

a.
b.
c.
d.

4. A line goes through the point (-4, 0) and the point (0,2).

Quantity A	Quantity B
The slope of the line	$\frac{1}{2}$

a.
b.
c.
d.

5.

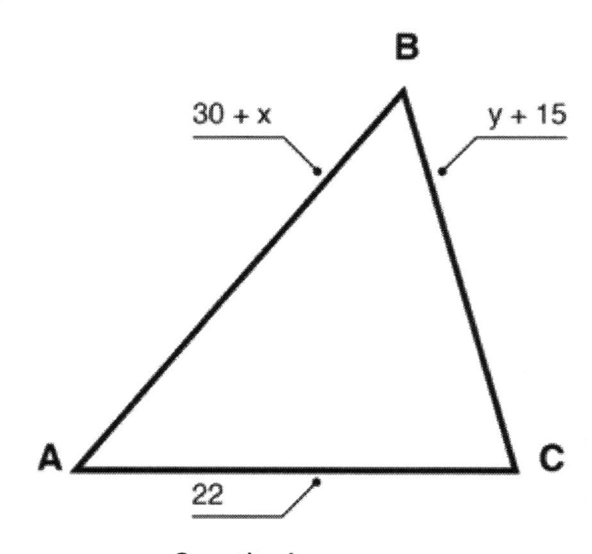

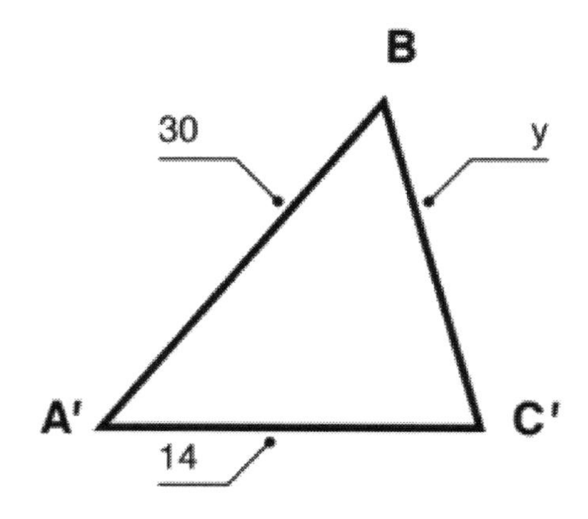

Quantity A	Quantity B
The value of y	26

a.
b.
c.
d.

6. $x^2 - 2xy + 2y^2$

Quantity A	Quantity B
The value of the solution when $x = 2, y = 3$	10

a.
b.
c.
d.

7. A sample data set contains the following values: 1, 3, 5, 7.

Quantity A	Quantity B
The standard deviation of the set	2.5

a.
b.
c.
d.

8. Two cards are drawn from a shuffled deck of 52 cards.

Quantity A	Quantity B
The probability that both cards are Kings if the first card isn't replaced after it's drawn	$\frac{1}{169}$

a.
b.
c.
d.

Questions 9-20 have several different formats. Unless otherwise directed, select a single answer choice for each question. For Numeric Entry questions, follow the instructions below.

Numeric Entry Questions

Enter your answer in the box(es) below the question.

1. Your answer may be an integer, a decimal, a fraction, and it may be negative.

2. If a question asks for a fraction, there will be two boxes. One is for the numerator and one is for the denominator.

3. Equivalent forms of the value, such as 1.5 and 1.50, are all correct. Fractions do not need to be reduced to lowest terms.

4. Enter the exact answer unless your question asks you to round your answer.

9. What is the interquartile range (IQR) of the following data set?
1, 4, 6, 6, 9, 10, 12, 17, 18
   a. 10
   b. 9
   c. 14.5
   d. 5
   e. 9.5

For the following question, select all that apply:

$$G = .035O + .26$$

10. The linear regression model above is based on an analysis of the price of a gallon of gas (G) at 15 gas stations compared to the price of a barrel of oil (O) at the time. Based on this model, which of the following statements are true?
   a. There is a negative correlation between G and O.
   b. When oil is $55 per barrel then gas is approximately $2.19 per gallon.
   c. The slope of the line indicates that as O increases by 1, G increases by .035.
   d. If the price of oil increases by $8 per barrel then the price of gas will increase by approximately $0.18 per gallon.
   e. If the price of gas is $4 per gallon, the price of a barrel of oil would be $105.50 per barrel.

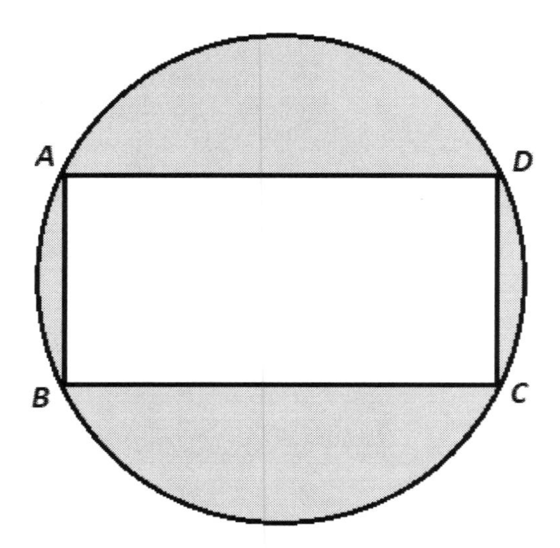

11. Rectangle ABCD is inscribed in the circle above. The length of side AB is 9 inches and the length of side BC is 12 inches. What is the area of the shaded region?

   a. 64.4 sq. in.
   b. 68.6 sq. in.
   c. 62.8 sq. in.
   d. 61.3 sq. in.
   e. 64.6 sq. in.

12. Given the value of a given stock at monthly intervals, which graph should be used to best represent the trend of the stock?

   a. Box plot
   b. Line plot
   c. Line graph
   d. Circle graph
   e. Dot plot

13. The graph shows the position of a car over a 10-second time interval. Which of the following is the correct interpretation of the graph for the interval 1 to 3 seconds?

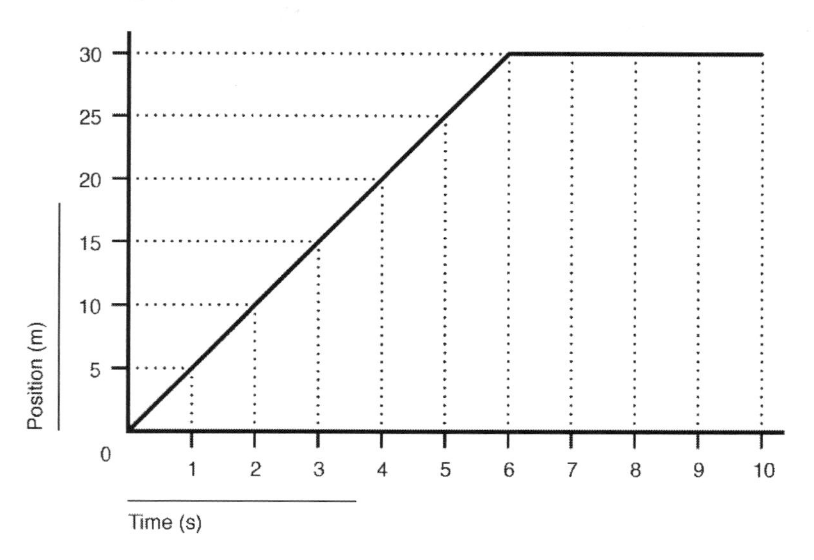

a. The car remains in the same position.
b. The car is traveling at a speed of 5m/s.
c. The car is traveling up a hill.
d. The car is traveling at 5mph.
e. The car accelerates at a rate of 5m/s.

For the following question, enter your answer in the box:

14. In Jim's school, there are 3 girls for every 2 boys. There are 650 students in total. Using this information, how many students are girls?

15. What type of function is modeled by the values in the following table?

X	f(x)
1	2
2	4
3	8
4	16
5	32

a. Linear
b. Exponential
c. Quadratic
d. Cubic
e. Logarithmic

16. Four people split a bill. The first person pays for $\frac{1}{5}$, the second person pays for $\frac{1}{4}$, and the third person pays for $\frac{1}{3}$. What fraction of the bill does the fourth person pay?

a. $\frac{1}{12}$

b. $\frac{47}{60}$

c. $\frac{1}{4}$

d. $\frac{4}{15}$

e. $\frac{13}{60}$

For the following question, enter your answer in the box:

17. What's the probability of rolling a 6 at least once in two rolls of a die?

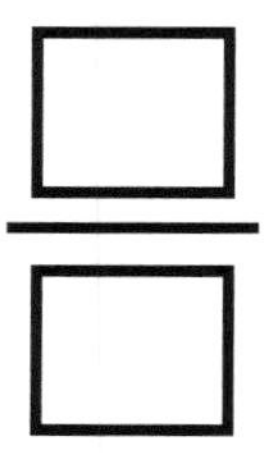

18. If the volume of a sphere is 288π cubic meters, what are the radius and surface area of the same sphere?

a. Radius 6 meters and surface area 144π square meters
b. Radius 36 meters and surface area 144π square meters
c. Radius 6 meters and surface area 12π square meters
d. Radius 36 meters and surface area 12π square meters
e. Radius 12 meters and surface area 144π square meters

19. How could the following equation be factored to find the zeros?

$$y = x^3 - 3x^2 - 4x$$

a. $0 = x^2(x - 4), x = 0, 4$
b. $0 = 3x(x + 1)(x + 4), x = 0, -1, -4$
c. $0 = x(x + 1)(x + 6), x = 0, -1, -6$
d. $0 = 3x(x + 1)(x - 4), x = 0, 1, -4$
e. $0 = x(x + 1)(x - 4), x = 0, -1, 4$

For the following question, enter your answer in the box:

20. Last year, the New York City area received approximately $27\,^3/_4$ inches of snow. The Denver area received approximately 3 times as much snow as New York City. How much snow fell in Denver?

Answer Explanations Test #2

1. A: 85% of a number means multiplying that number by 0.85. So, $0.85 \times 20 = \frac{85}{100} \times \frac{20}{1}$, which can be simplified to $\frac{17}{20} \times \frac{20}{1} = 17$. Since 17 is greater than 16 (the value of *Quantity B*), the value of *Quantity A* is larger, so Choice *A* is correct.

2. B: If the average of all six numbers is 6, that means $\frac{a+b+c+d+e+x}{6} = 6$. The sum of the first five numbers is 25, so this equation can be simplified to $\frac{25+x}{6} = 6$. Multiplying both sides by 6 gives $25 + x = 36$, and x, or the sixth number, is found to equal 11. This means that *Quantity B* (12) is greater than *Quantity A*.

3. A: $\frac{5}{2} \div \frac{1}{3} = \frac{5}{2} \times \frac{3}{1} = \frac{15}{2} = 7.5$. This means that *Quantity A* is greater than *Quantity B.*

4. C: The slope is given by the change in *y* divided by the change in *x*. The change in *y* is 2-0 = 2, and the change in *x* is 0 – (-4) = 4. The slope is $\frac{2}{4} = \frac{1}{2}$. This means that *Quantity A* is equal to *Quantity B.*

5. A: Because the triangles are similar, the lengths of the corresponding sides are proportional. Therefore:

$$\frac{30 + x}{30} = \frac{22}{14} = \frac{y + 5}{y}$$

This results in the equation $14(30 + x) = 22 \cdot 30$ which, when solved, gives $x = 17.1$. The proportion also results in the equation $14(y + 5) = 22y$ which, when solved, gives $y = 26.3$. Since 26.3 is greater than 26, *Quantity A* is greater than *Quantity B.*

6. C: Each instance of *x* is replaced with a 2, and each instance of *y* is replaced with a 3 to get $2^2 - 2 \cdot 2 \cdot 3 + 2 \cdot 3^2 = 4 - 12 + 18 = 10$. This means the *Quantity A* is equal to *Quantity B.*

7. A: First, the sample mean must be calculated. $\bar{x} = \frac{1}{4}(1 + 3 + 5 + 7) = 4$. The standard deviation of the data set is $\sigma = \sqrt{\frac{\Sigma(x-\bar{x})^2}{n-1}}$, and $n = 4$ represents the number of data points. Therefore:

$$\sigma = \sqrt{\frac{1}{3}[(1-4)^2 + (3-4)^2 + (5-4)^2 + (7-4)^2]} = \sqrt{\frac{1}{3}(9 + 1 + 1 + 9)} = 2.58$$

This is greater than the value in *Quantity B.* Even if rounded one decimal place, *Quantity A* (which would then become 2.6) is greater than *Quantity B.*

8. B: For the first card drawn, the probability of a King being pulled is $^4/_{52}$. Since this card isn't replaced, if a King is drawn first, the probability of a King being drawn second is $^3/_{51}$. The probability of a King being drawn in both the first and second draw is the product of the two probabilities: $^4/_{52} \times ^3/_{51} = ^{12}/_{2652}$ which, divided by 12, equals $^1/_{221}$. With fractions, if the numerator is the same but one denominator is larger than the other, the fraction with the smaller value in the denominator is larger. In this case, $^1/_{169}$ is larger than $^1/_{221}$. Therefore, *Quantity B* is greater than *Quantity A.*

9. E: The interquartile range (IQR) is the difference between the value of the 75th percentile and the 25th percentile of a data set, or the third and first quartiles. The numbers in the data set must first be arranged from lowest to highest value, as they are in this problem. When the number of data points in the set is odd (as it is in this problem), the median, which is the second quartile, is not included in the upper and lower quartiles. The remaining data points are then divided into two equal groups (in this case, the first four numbers comprise the first quartile, the middle number is excluded, and the largest four numbers make up the upper quartile). Then, the median of both quartiles is found, remembering that the median is the average of the two middle points if the number of values is even, as it is in this case, with four numbers. Finally, the IQR is calculated as the difference between the median of the upper quartile and that of the lower quartile.

So, for this problem, the data set provided was: 1, 4, 6, 6, 9, 10, 12, 17, 18.

There are nine values, so the median (9) is excluded. This makes the lower quartile is 1, 4, 6, and 6 and the upper quartile is 10, 12, 17, and 18. The median of the lower quartile is 5 (the average of 4 and 6) and the median of the upper quartile is 14.5 (the average of 12 and 17).

Thus, the IQR is 14.5 — 5 = 9.5.

10. B & C: The statements in Choices *B* and *C* are the only true statements. If graphed this line would have a positive correlation, which makes the statement in Choice *A* false. If the price of oil increases by \$8 per barrel then gas price would increase by $.035(8) = \$0.28$ per gallon. This makes the statement in Choice *D* false as well. Lastly, if the price of is \$9, then the price of oil would be:

1. $4 = .035O + .26$
2. $3.74 = .035O$
3. $O = \$106.86/barrel$

Therefore, the statement in Choice *E* is also incorrect.

11. B: The inscribed rectangle is 9 X 12 inches. First find the length of *AC* using the Pythagorean Theorem. So, $9^2 + 12^2 = c^2$, where *c* is the length of *AC* in this case. This means that *AC* = 15 inches. This means the diameter of the circle is 15 inches. This can be used to find the area of the entire circle. The formula is πr^2. So, $3.14(7.5)^2 = 176.6$ sq. inches. Then take the area of the rectangle away to find just the area of the shaded region. This is 176.6 – 108 = 68.6.

12. C: The scenario involves data consisting of two variables: month and stock value. Box plots display data consisting of values for one variable. Therefore, a box plot is not an appropriate choice. Both line plots (which are also called dot plots) and circle graphs are used to display frequencies within categorical data. Neither can be used for the given scenario. Line graphs display two numerical variables on a coordinate grid and show trends among the variables, so this is the correct choice.

13. B: The car is traveling at a speed of five meters per second. On the interval from one to three seconds, the position changes by fifteen meters. By making this change in position over time into a rate, the speed becomes ten meters in two seconds or five meters in one second.

13. 390: Three girls for every two boys can be expressed as a ratio: 3:2. This can be visualized as splitting the school into 5 groups: 3 girl groups and 2 boy groups. The number of students in each group can be found by dividing the total number of students by 5:

650 divided by 5 equals 1 part, or 130 students per group

To find the total number of girls, the number of students per group (130) is multiplied by the number of girl groups in the school (3). This equals 390.

15. B: The table shows values that are increasing exponentially. The differences between the inputs are the same, while the differences in the outputs are changing by a factor of 2. The values in the table can be modeled by the equation $f(x) = 2^x$.

16. E: To find the fraction of the bill that the first three people pay, the fractions need to be added, which means finding common denominator. The common denominator will be 60. $\frac{1}{5} + \frac{1}{4} + \frac{1}{3} = \frac{12}{60} + \frac{15}{60} + \frac{20}{60} = \frac{47}{60}$. The remainder of the bill is $1 - \frac{47}{60} = \frac{60}{60} - \frac{47}{60} = \frac{13}{60}$.

17. $11/36$: The addition rule is necessary to determine the probability because a 6 can be rolled on either roll of the die. The rule used is $P(A \text{ or } B) = P(A) + P(B) - P(A \text{ and } B)$. The probability of a 6 being individually rolled is $^1/_6$ and the probability of a 6 being rolled twice is $^1/_6 \cdot {^1/_6} = {^1/_{36}}$. Therefore, the probability that a 6 is rolled at least once is $1/6 + 1/6 - 1/36 = 11/36$

18. A: Because the volume of the given sphere is 288π cubic meters, this gives $^4/_3\, \pi r^3 = 288\pi$. This equation is solved for r to obtain a radius of 6 meters. The surface area of a sphere is $4\pi r^2$ so, if $r = 6$ in this formula, the surface area is 144π square meters.

19. E: Finding the zeros for a function by factoring is done by setting the equation equal to zero, then completely factoring. Since there was a common x for each term in the provided equation, that is factored out first. Then the quadratic that is left can be factored into two binomials: $(x + 1)(x - 4)$. Setting each factor equation equal to zero and solving for x yields three zeros.

20. 83 $^1/_4$ inches: 3 must be multiplied times $27\, ^3/_4$. In order to easily do this, the mixed number should be converted into an improper fraction. $27\, ^3/_4 = {}^{27*4+3}/_4 = {}^{111}/_4$. Therefore, Denver had approximately $3 \times {}^{111}/_4 = {}^{333}/_4$ inches of snow. The improper fraction can be converted back into a mixed number through division. ${}^{333}/_4 = 83\, ^1/_4$ inches.

Dear GRE Test Taker,

We would like to start by thanking you for purchasing this study guide for the math section of your GRE Exam. We hope that we exceeded your expectations.

Our goal in creating this study guide was to cover all of the topics that you will see on the math section. We also strove to make our practice questions as similar as possible to what you will encounter on test day. With that being said, if you found something that you feel was not up to your standards, please send us an email and let us know.

We would also like to let you know about other books in our catalog that may interest you.

GRE (for all sections)

This can be found on Amazon: amazon.com/dp/1628455004

GMAT

amazon.com/dp/1628454539

MCAT

amazon.com/dp/1628455012

We have study guides in a wide variety of fields. If the one you are looking for isn't listed above, then try searching for it on Amazon or send us an email.

Thanks Again and Happy Testing!
Product Development Team
info@studyguideteam.com

FREE Test Taking Tips DVD Offer

To help us better serve you, we have developed a Test Taking Tips DVD that we would like to give you for FREE. **This DVD covers world-class test taking tips that you can use to be even more successful when you are taking your test.**

All that we ask is that you email us your feedback about your study guide. Please let us know what you thought about it – whether that is good, bad or indifferent.

To get your **FREE Test Taking Tips DVD**, email freedvd@studyguideteam.com with "FREE DVD" in the subject line and the following information in the body of the email:

a. The title of your study guide.

b. Your product rating on a scale of 1-5, with 5 being the highest rating.

c. Your feedback about the study guide. What did you think of it?

d. Your full name and shipping address to send your free DVD.

If you have any questions or concerns, please don't hesitate to contact us at freedvd@studyguideteam.com.

Thanks again!

Made in the USA
San Bernardino, CA
25 June 2018